1
NEIGHBORHOOD FACILITY

MINI
Building

CONTENTS

MINI Building

NEIGHBORHOOD FACILITY

SUGAMO SHINKIN BANK / TOKIWADAI BRANCH

스가모 신용금고/토키와다이 지점

We handled the architectural and interior design for the bank's newly relocated branch in Tokiwadai. By basing our design around leaf motifs, we sought to create a refreshing space. The facade of the building features silhouettes of trees and an assortment of both large and small windows in 14 different colors arranged in a distinctive, rhythmical pattern that transforms the facade itself into signage. ATMs and teller windows are located on the first floor, along with 3 courtyards and an open space laid out with chairs in 14 different colors. The second storey houses the loan section, reception rooms, offices and 4 courtyards, while the third floor is reserved for facilities for staff use, including changing rooms and a cafeteria. Thanks to the 7 light-filled courtyards planted with trees and flowering plants, each of these spaces is loosely connected to all the others. A constellation of leaves in 24 different colors growing on the white branches of the walls and glass windows overlaps with the natural foliage of the real trees in the courtyards, creating the sensation of being in a magical forest.

본 프로젝트는 토키와다이로 이전한 스가모 신용금고의 신축과 내부 공간 디자인으로, 나뭇잎 모티브를 토대로 편안하면서 생기가 넘치는 분위기의 공간을 연출하고자 했다. 입면은 나무 실루엣을 이용한 패턴과 다양한 색상과 크기의 창을 계획하여 색다르고 리드미컬한 특징을 부여했다. ATM과 창구는 세 개의 중정, 개방된 공간과 함께 1층에 배치했다. 2층에는 대출, 대기 공간, 업무 공간 그리고 4개의 중정이 있으며 3층에는 휴게실 및 카페테리아와 같은 직원 전용 공간을 배치했다. 나무와 꽃이 심어진 중정은 다른 공간과 긴밀하게 소통한다. 벽과 유리창의 채색된 나무 패턴은 중정 안에 심어진 실제 나무들과 어우러지면서 신비로운 분위기를 연출한다.

Architect : emmanuelle moureaux architecture + design
Location : Tokiwadai, Tokyo, Japan **Completion :** 2010. 6
Photographer : Nacasa & Partners Inc.

설계 : 엠마뉴엘 무로 아키텍처 + 디자인 **위치 :** 일본 도쿄 토키와다이 **완공 :** 2010. 6 **사진 :** 나카사 & 파트너스

← View from North

↑ View of the main facade

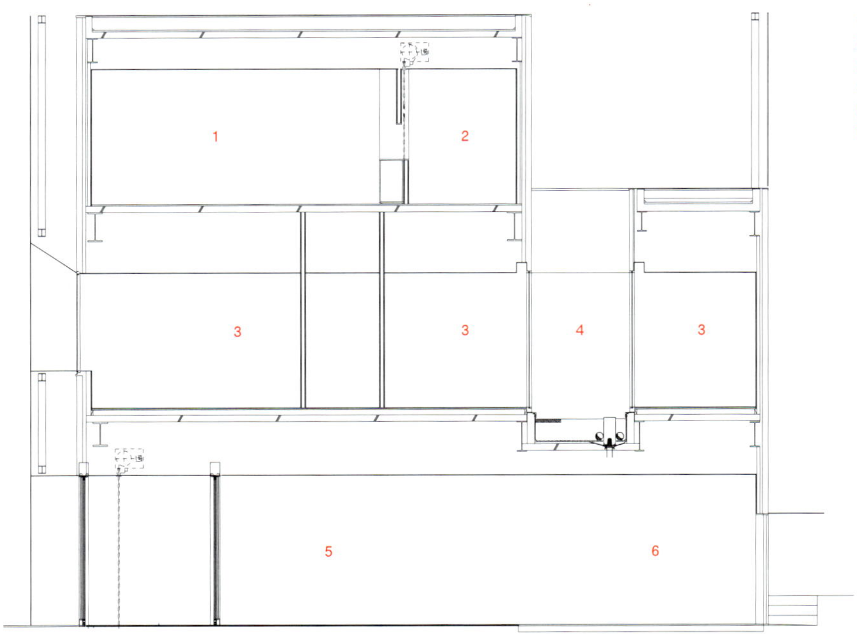

1 CAFETERIA
2 KITCHEN
3 MEETING SPACE
4 GARDEN
5 OPEN SPACE
6 OFFICE

SECTION

↰ View of the open space on the 1st floor ↑ Inner courtyards on the 2nd floor ↓ View of the open space with inner garden

↰ Night view from West
↲ View of the open space from outside
→ Entrance

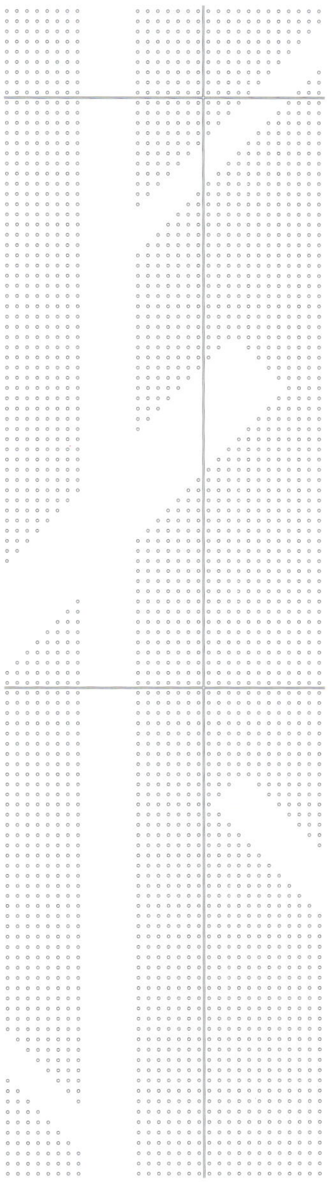

FACADE

View of the office area having courtyards
Stairs have void space
Courtyard next to office area

1 OFFICE
2 OPEN SPACE
3 GARDEN
4 MEETING SPACE
5 ARCHIVE

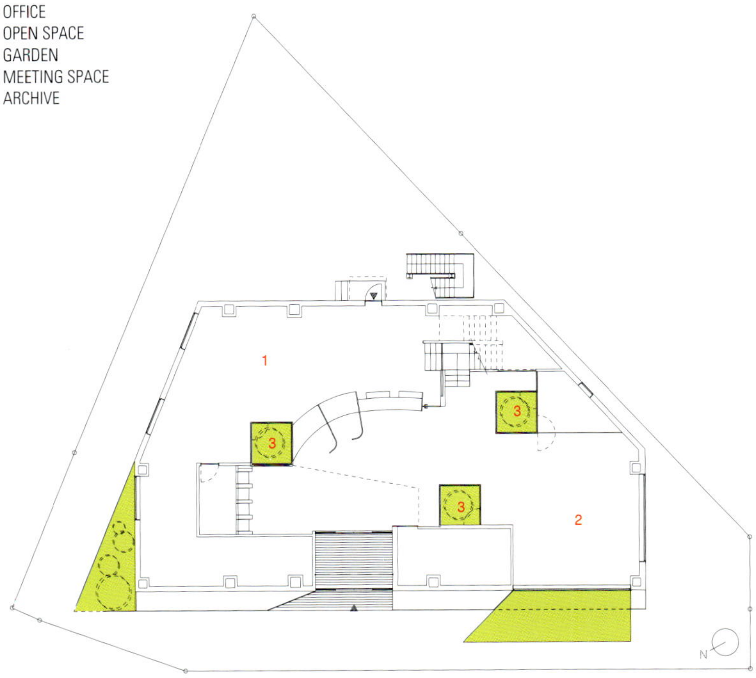

1ST FLOOR PLAN

2ND FLOOR PLAN

CAFE MO' BETTER BLUES

동숭동 모 베터 블루스

Cafe Mo' Better Blues is located in a quite place behind the busy commercial area in Daehak-ro. Its original building was a 12-year old one-story cafe. The client wanted a new building to have scenery and memory of the old cafe. In that situation, we tried to plan the building as a device to recall shabby and old rural scenery in the city. To secure enough space, the area of stairs to connect each floor had to be minimized. Therefore the stairs to connect the 1st and 2nd floors had a cantilever structure and the stairs to connect from the 2nd floor to the 5th floor were planned to have 75cm-wide width. Windows in various sizes that show outside views were installed to resolve narrowness of the space. Materials for the outer wall were exposed concrete and mortar. With new interpretation of cement mortar, we intended to express scenery of the old cafe. Terraces were arranged in different locations on each floor. They are another device to recall old scenery. Plants and trees on those green terraces in the rear of the 1st floor, front of the 2nd floor, rear and right side of the 3rd floor, front of the 4th floor and the yard of the 5th floor will decorate the whole building and hold traces of times. Especially the terrace in the rear of the 3rd floor was constructed in a cantilever structure to express the non-rectangular shape of the site.

카페 모 베터 블루스는 대학로의 번잡한 상업시설 가로 뒤편의 조용한 곳에 위치한다. 건축주는 12년이나 된 1층짜리 기존 카페의 풍경을 이어나갈 수 있는 건물을 원했다. 따라서 대도시 속에서 낡고 오래된 시골의 풍경을 기억할 수 있는 장치로 계획하고자 했다. 각 층을 연결하는 계단의 면적을 최소화해야만 사용 가능한 면적을 얻을 수 있었다. 1층과 2층을 연결하는 계단은 캔틸레버 구조로 하여 면적을 최소화했고, 2층에서 5층을 연결하는 계단은 폭을 75cm로 계획했다. 계단 주변 곳곳에 외부를 볼 수 있는 창문을 다양한 크기로 설치하여 공간의 협소함을 해소시켰다. 외벽은 노출콘크리트(미송소폭판)와 시멘트 모르타르로 마감했다. 시멘트 모르타르를 새롭게 해석하여 오래된 카페의 풍경을 담아내고자 했다. 테라스는 각 층마다 각기 다른 위치에 테라스를 배치했다. 테라스 또한 풍경을 기억하게 하는 장치 중 하나이다. 1층 후면, 2층의 전면, 3층의 후면과 우측, 4층의 전면, 5층의 마당 등 각 층마다 크고 작은 테라스에 화분을 놓고 나무를 심어 건물 전체를 감싸며 세월의 흔적을 간직하도록 했다. 특히 3층 후면의 테라스는 부정형인 대지의 형상을 담아내고자 캔틸레버 구조로 설치했다.

Architect : modo architect office(Kim Kyung Hee + Lee Eung Rak) **Location :** 26 Dongsung-dong, Jongno-gu, Seoul, Korea **Site area :** 106.51m² **Building area :** 63.90m² **Gross floor area :** 276.46m² **Building coverage :** 59.99% **Floor space index :** 199.52% **Design period :** 2008. 8 ~ 2008. 12 **Construction period :** 2009. 2 ~ 2009. 10 **Photographer :** Yum Seung Hoon

설계 : 건축사사무소 모도건축(김경희 + 이응락) **위치 :** 서울특별시 종로구 동숭동 26번지 외 1필지 **대지면적 :** 106.51㎡ **건축면적 :** 63.90㎡ **연면적 :** 276.46㎡ **건폐율 :** 59.99% **용적률 :** 199.52% **설계기간 :** 2008. 8 ~ 2008. 12 **공사기간 :** 2009. 2 ~ 2009. 10 **사진 :** 염승훈

← View of the front left side → Former building

View of the front right side Entrance

FRONT ELEVATION

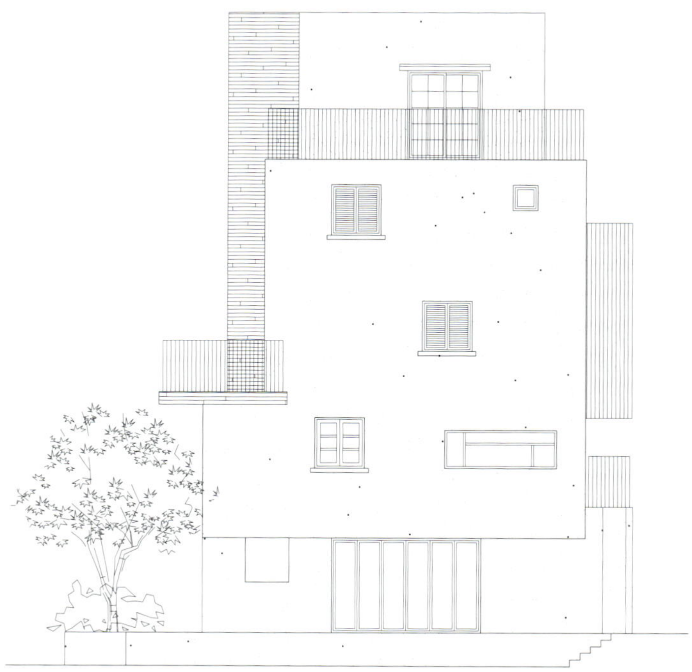

LEFT ELEVATION

↓ ↑ Interior view of the 3rd floor

cafe
mo'better
blues
since
1997

← View of the facade ↱ Stairs of the 3rd floor ↳ Stairs of the 4th floor

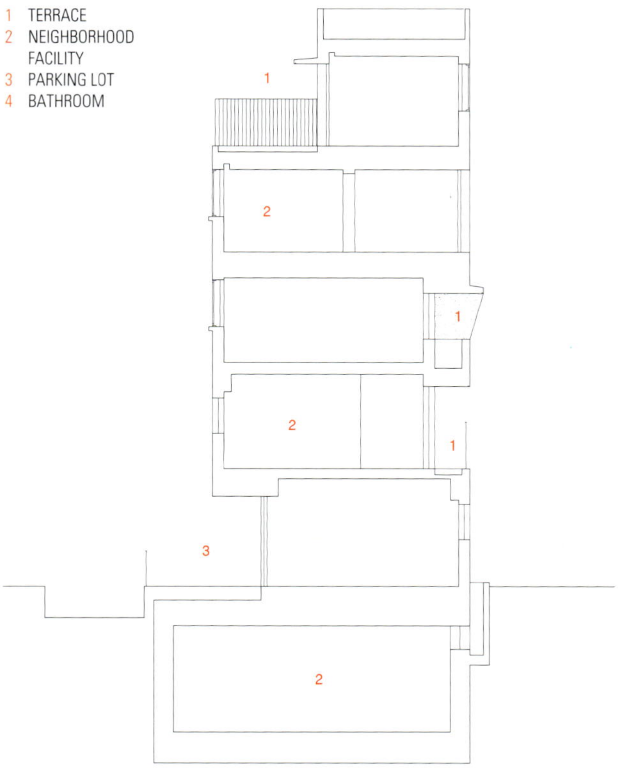

CROSS SECTION

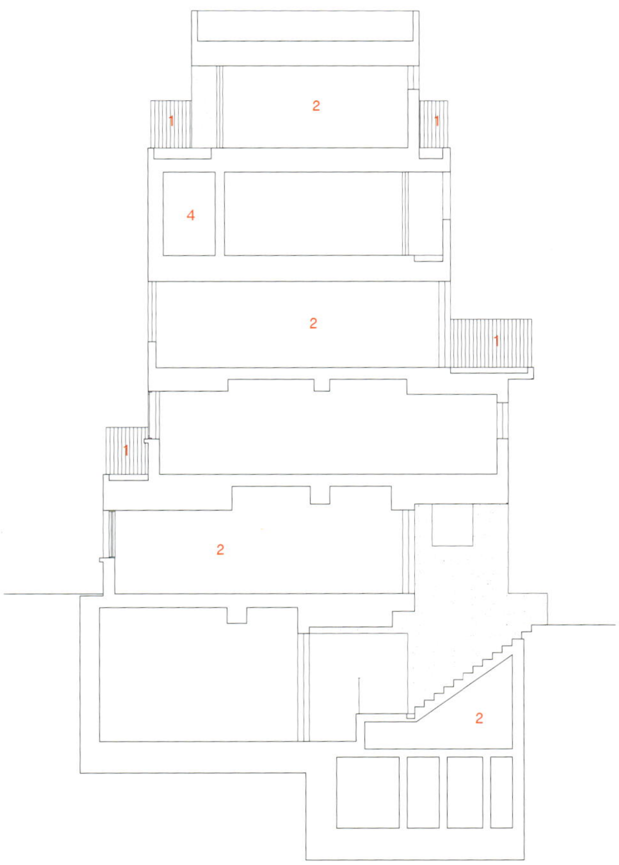

LONGITUDINAL SECTION

← Stairs of the 4th floor ↰ Roof garden ↱ Overall view of right side

1 6M ROAD
2 3M ROAD
3 PARKING LOT
4 NEIGHBORHOOD FACILITY
5 TERRACE

1ST FLOOR PLAN

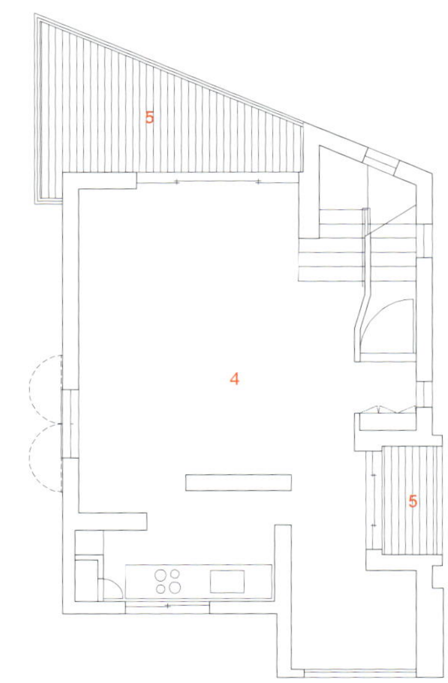

3RD FLOOR PLAN

HARMONIA 57

하모니아 57

Like a living body, the building breathes, sweats and modifies itself, transcending its inertia. The walls are thick and covered externally by a vegetal layer that works like the skin of the structure. This dense wall is made of an organic concrete that has pores, where several plant species grow, giving the facades a unique look. Its insides are exposed in the facades while the interior spaces are well finished with clear and luminous surfaces, as if the construction was inside out. The pipelines that serve the whole building - as well as the pumps and the water treatment system - are showing in the exterior walls, embracing them like veins and arteries of a body. Its volume is quite simple, but also remarkable: two grand vegetal blocks are connected by a metallic footbridge, cut by concrete and glass windows and terraces. Between the blocks a internal plaza opens like a clearing and acts like a place of encountering. The terraces are spread in each floor, creating a visual game between volumes, lighting and transparency in the internal spaces. The frontal block is completely suspense, levitating over pilotis, while the back block is solid, complemented by a birdhouse-like volume on top of it.

Architect : TRIPTYQUE(Greg Bousquet, Carolina Bueno, Guillaume Sibaude, Olivier Raffaelli) **Location :** São Paulo, Brazil **Surface :** 1,100m² **Completion :** 2008 **Photographer :** Leonardo Finotti, Nelson Kon

설계 : 트립티크(그레그 부스케, 캐롤리나 부에노, 쥴리아움 지보, 올리비에 라파엘리) **위치 :** 브라질 상파울루 **면적 :** 1,100㎡ **완공 :** 2008 **사진 :** 레오나르두 피노티, 넬슨 콘

살아있는 유기체처럼 건물은 호흡하고, 땀을 흘리고 스스로 변신한다. 식물층으로 외부가 완전히 덮인 두꺼운 벽들은 구조체의 외피 역할을 한다. 벽체에는 몇 가지 식물들이 자랄 수 있도록 구멍이 있어 건물 파사드로써 독특한 외관을 조성한다. 파사드로 노출되는 내부 공간들은 마치 외부로 보이기 위한 것처럼 깨끗하고 광택이 나는 표면으로 마감했다. 건물 전체에는 펌프 및 배수 시스템이 파이프 선으로 뒤덮여 있어 마치 외벽을 정맥과 동맥이 감싼 듯 노출되어있다. 건물의 볼륨은 아주 심플하지만 동시에 두드러진다. 식물들로 덮인 두 개의 커다란 블록은 콘크리트와 유리창, 테라스 공간으로 구분되며 금속재질의 다리로 연결된다. 블록들 사이에 자리한 내부 플라자는 시원하게 개방되어 있어 만남의 장으로서의 역할을 한다. 각 층에 펼쳐져 있는 테라스는 내부 공간의 볼륨, 조명, 투명성 사이에서 하나의 게임을 창출한다. 후면 블록이 새집처럼 생긴 볼륨에 의해 보강된 견고한 구조라면 정면의 블록은 완전히 필로티 위로 허공에 떠 있다.

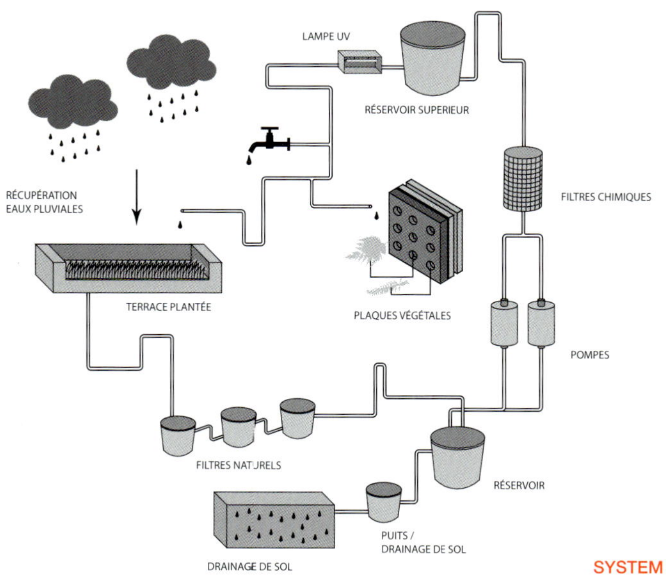

SYSTEM DIAGRAM

← View from the deck of the 1st floor

Front facade Overall view showing the urban context Exterior view from the side

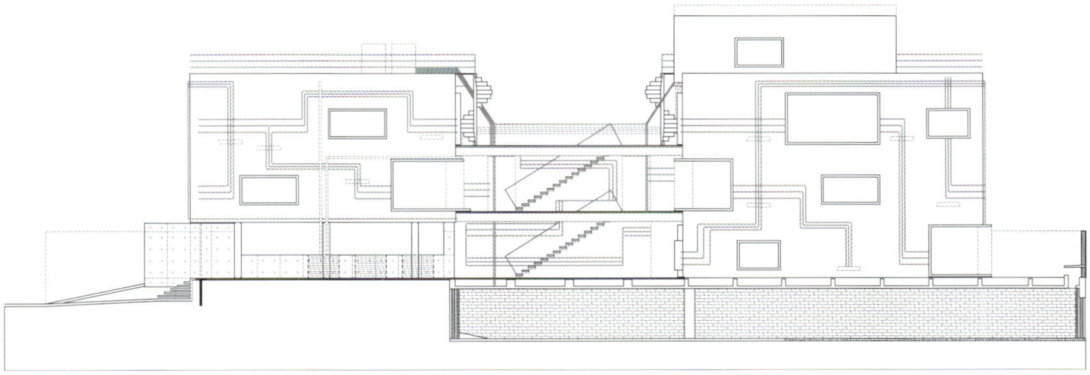

ELEVATION

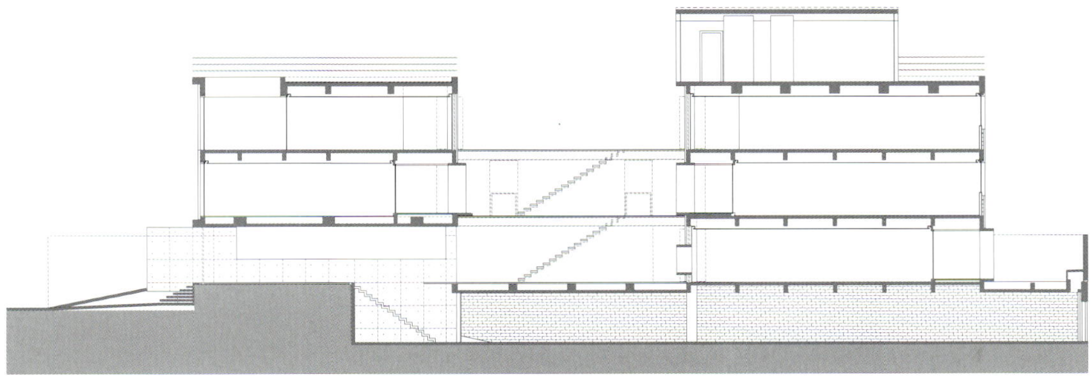

LONGITUDINAL SECTION

Two volumes with internal plaza

← View from the deck of the 1st floor ↰ External stairs and footbridge ↱ View from the deck of the 2nd floor

1 INTERNAL PLAZA
2 FOOTBRIDGE
3 ENTRANCE

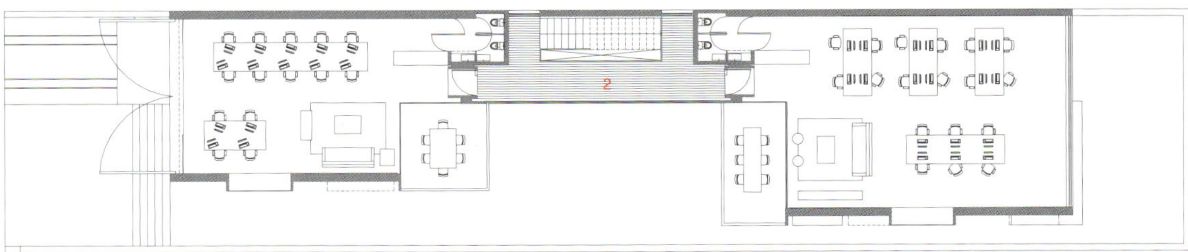

1ST FLOOR PLAN

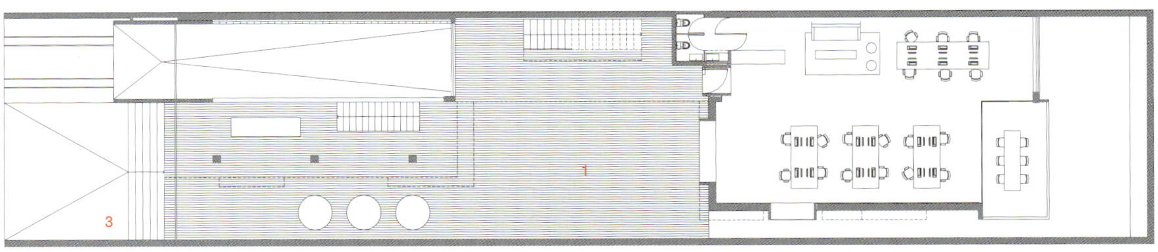

GROUND FLOOR PLAN

SLIT COURT

슬릿 코트

The essence of this architecture lies in an "inner court". This building is a collective housing of five stories. The site of a Japanese house is generally small and a narrow lot is facing to a narrow street. Even in a historic town, it happens that such a narrow street is turned to a main street to be the transportation route for cars, then it is no more a place where town people had once shared joy of living together. Therefore in our architecture, we make the inner living space a place to protect what matters in life. We try to protect sky, land and light "inside the building." This inner court redefines how the form of "people, town and path" should be in Kyoto. Organic curved slits winding from the ground to the sky are carved on the surface of four concrete walls which enclose the inner court. The slits do not all have the same width, but are made to fluctuate from thick to thin and widen again as the curves of the slits rise. The walls on which the slits are carved are 200mm thick. To make the construction of the concrete framework and cutting of the glass simpler, all the curves are composed of a series of circular arcs. Designing circular arcs that express an organic, pulsing feeling was an extremely delicate and sensitive work. The facade consists of a lower part with its cherry tree motif, and an upper part, with graded louvers arranged according to pitch. The lower and upper part is clearly divided with a 760mm wide space. The lower part composed of curved slits which form a series of circular arcs. Like a tree, it evokes a sense of energy gathering at the center.

슬릿 코트의 본질은 "안뜰"에 있다. 이 건물은 5층 짜리 공동 주택으로, 일반적인 일본의 소규모 대지와 같이 협소하며 좁은 거리에 면해 있다. 좁은 골목길이 편리를 위해 자동차가 통행하는 대로로 사용되면서 유서 깊은 이 마을에 더이상 사람들이 부대끼며 사는 기쁨을 나눌 만한 장소를 찾아보기 어렵게 되었다. 슬릿 코트 프로젝트를 통하여 인생의 중요한 것들을 지켜줄 수 있는 내부의 삶의 공간을 만들고자 했다. 우리는 하늘, 땅 그리고 햇빛을 "건물 안에서" 지켜내고자 한다. 이 안뜰은 "사람들, 도시 그리고 길목"의 형태가 교토에서 어떻게 이루어져야 하는지를 재정의한다. 지면에서 하늘로 이어지는 유기적인 곡선형 슬릿이 안뜰을 감싸는 네 개의 콘크리트 벽 표면에 새겨진다. 슬릿의 폭은 각각 다르다. 폭이 좁아 졌다가 넓어지는 변동을 거듭하지만 상승하는 부분에서 곡선이 다시 넓어진다. 콘크리트 작업과 유리 재단이 용이하도록 모든 곡선은 일련의 원호(arc)로 구성된다. 전면 파사드는 체리나무를 모티브로 한 하부와 나란하게 배열된 루버 상부로 이루어진다. 하부와 상부는 760mm 너비 공간으로 분명하게 구분된다. 하부는 원호를 형성하는 곡선 슬릿들로 이루어지고, 마치 나무와 같이 파사드의 중심으로 에너지가 모이는 형상을 나타낸다.

Architect : EASTERN design office(Anna Nakamura, Taiyo Jinno) **Location :** Kyoto, Japan **Site area :** 440.12m² **Gross floor area :** 859.42m² **Photographer :** Koichi Torimura

설계 : 이스턴 디자인 오피스(안나 나카무라, 타이요 진노) **위치 :** 일본 교토 **대지면적 :** 440.12㎡ **연면적 :** 859.42㎡ **사진 :** 코이치 토리무라

← View of the main facade → Courtyard

Facade of courtyard Main entrance

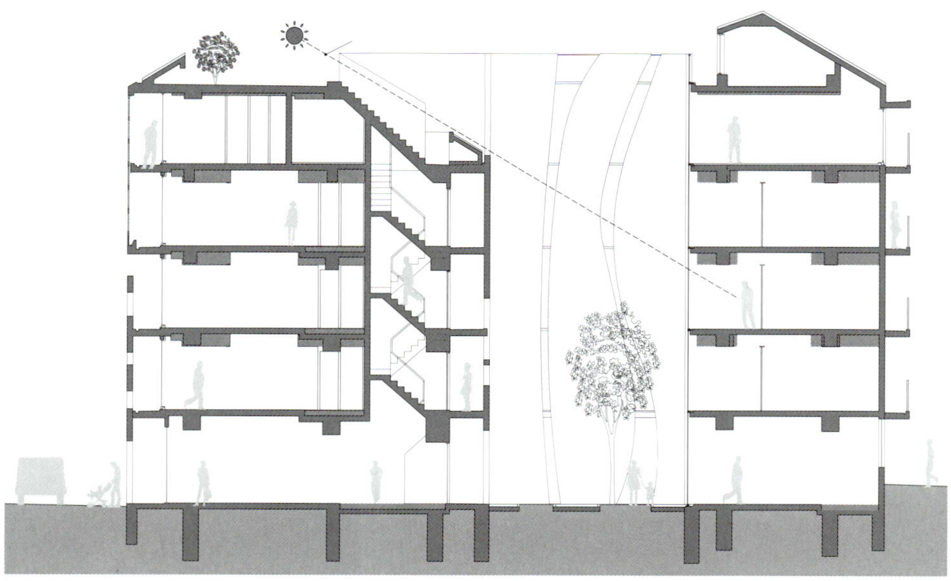

SECTION

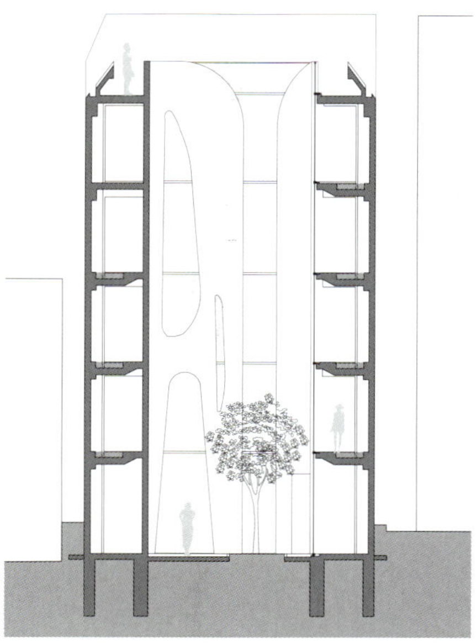

SECTION

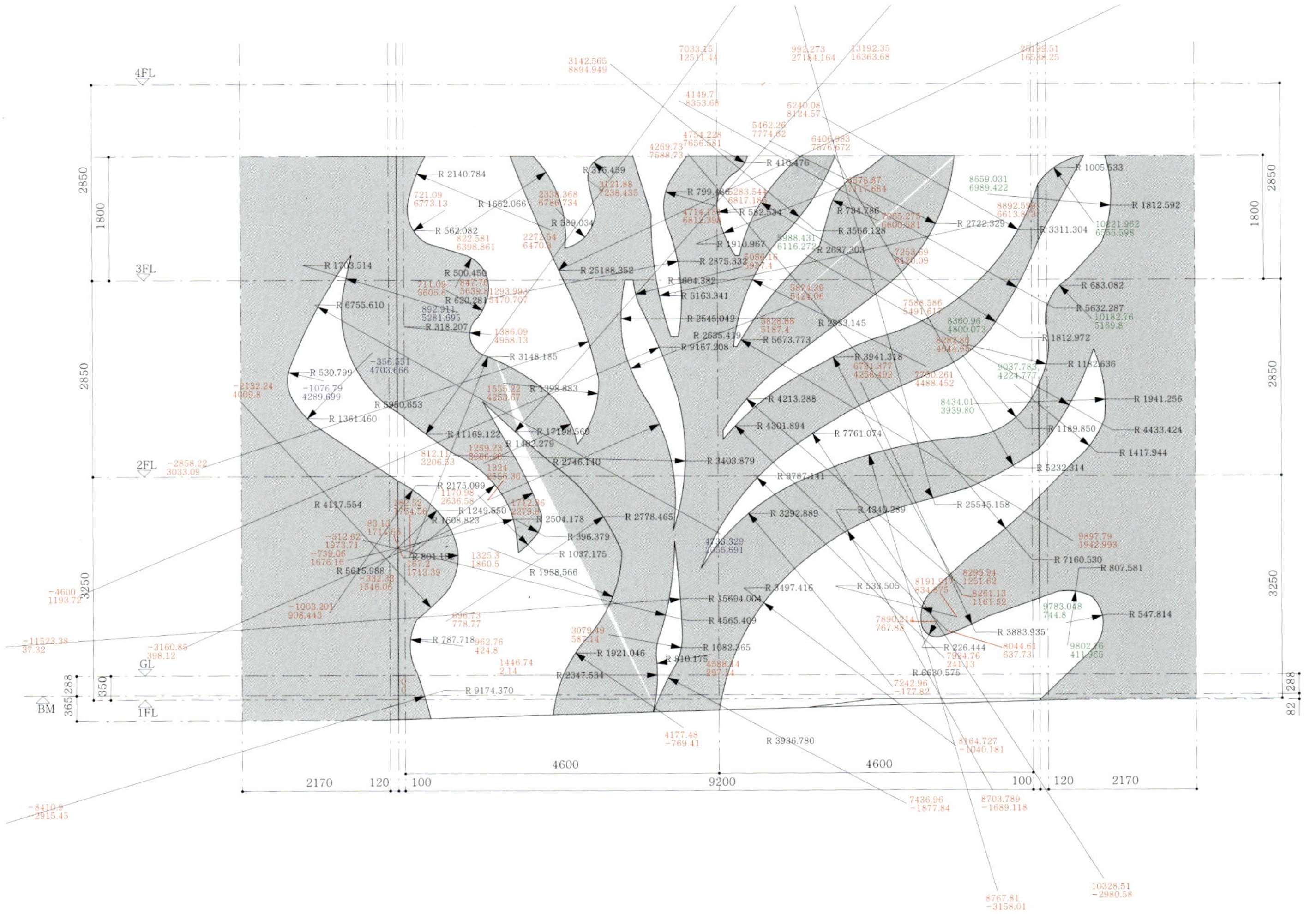

MAIN FACADE DEVELOPMENT FIGURE

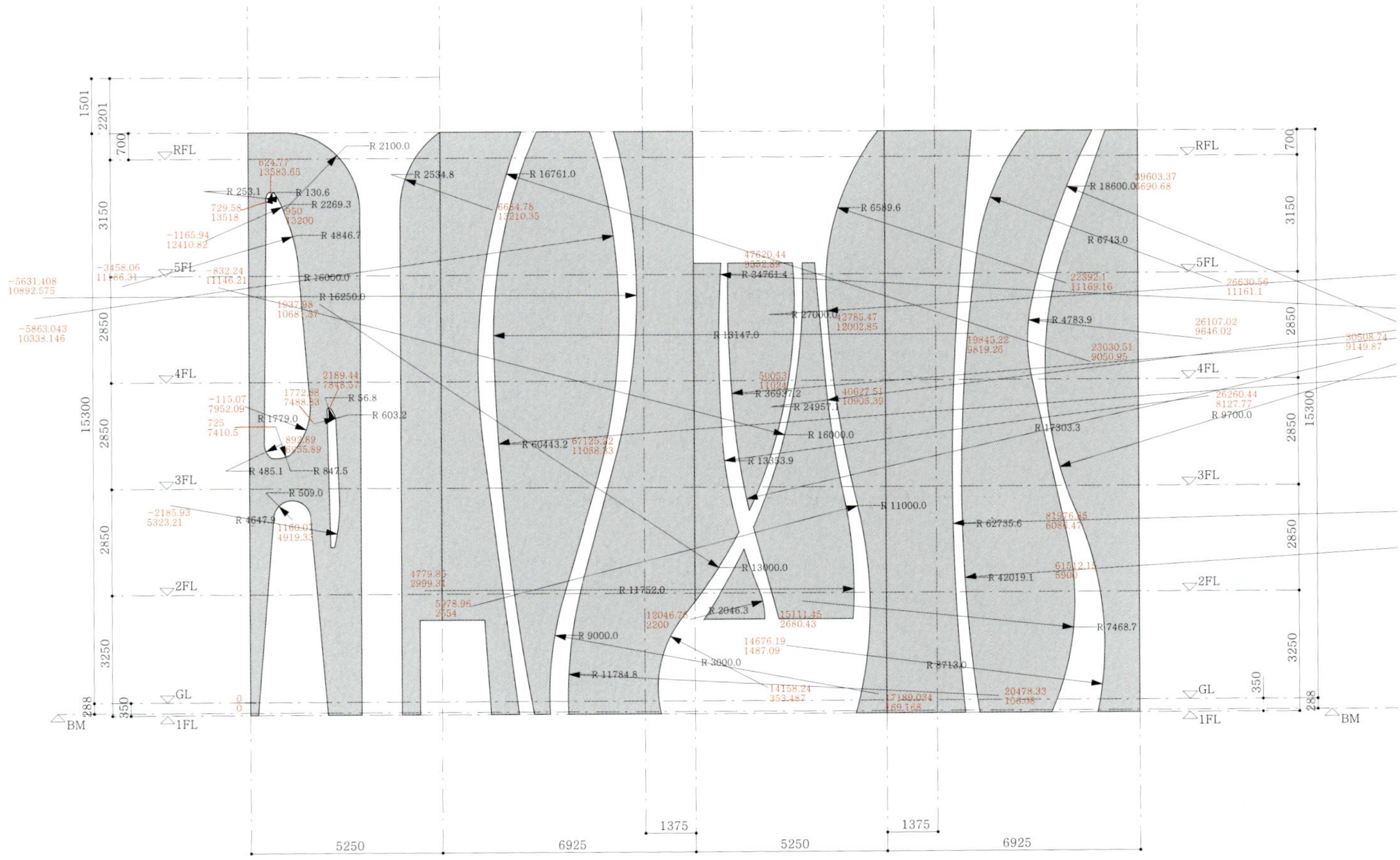

INNER COURT FACADE DEVELOPMENT FIGURE

← Top-down view of courtyard ↰ View from inside ↱ Interior view

1 TENANT SPACE
2 COURTYARD
3 HOUSING SPACE
4 TERRACE
5 STORE

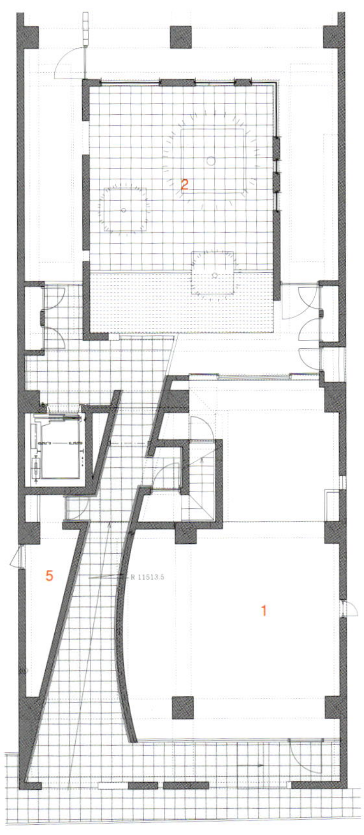

1ST FLOOR PLAN

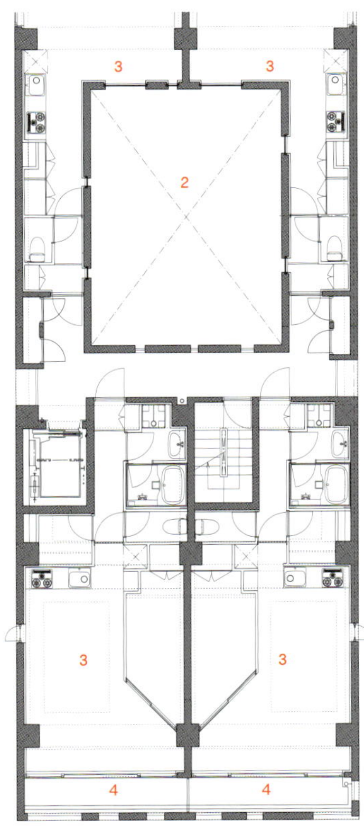

2ND FLOOR PLAN

GRANDE
OUVERTURE
ABONNEZ-VOUS
Mort de peine
845-7277
Pins
100

THÉÂTRE DE QUAT'SOUS

콰수스 극장

Serious problems related to the security and comfort of the users combined with the need for production spaces led to the decision to demolish and rebuild Theatre de Quat'sous on it's actual site after many feasibility studies demonstrated the impossibility to renovate the crumbling wood and brick structure built in 1907 as three row houses. Theater is about fugacity, a succession of unique moments that barely survive in the memories of those who were there. Theaters are ghostly figures that have witnessed what we are about to forget. In reconstructing the Theatre de Quat'sous we were specifically asked to incorporate whatever we could from the original building to help the new one evoke those memories. We chose to sample textures, images, colors and materials from a cultural inventory of the theater and mapped them on the assemblage of required volumes(stage, house, foyer, crossover, control booth and rehearsal). Recycling on site stones, slate, wood, bricks, marble and furniture becomes part of a strategy of cultural sustainability. New materials ncludesilkscreened glass, black brick and perforated aluminum that contribute to make Theatre de Quat'sous a ghostly figure accumulating memories.

Architect : Les architectes FABG(Eric Gauthier) **Location :** Montreal, Quebec, Canada **Area :** 860m² **Completion :** 2009. 4 **Photographer :** Steve Montpetit

설계 : 레 아키텍츠 FABG(에릭 고티에) **위치** : 캐나다 퀘벡 몬트리올 **면적** : 860m² **완공** : 2009. 4 **사진** : 스티브 몬피티

Existing building

Red stairway to balcony with framed pictures

Decorative steel door

Windows in the theater

Slate roof

Window lintel with floral pattern

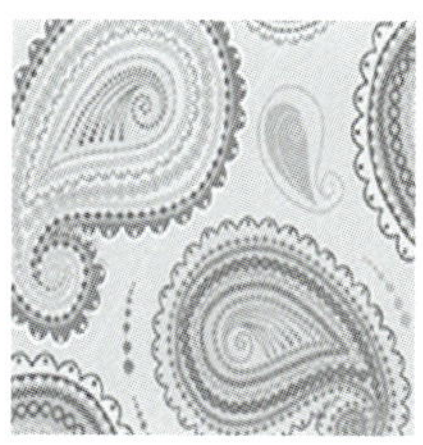
Paisley motif of the 60's

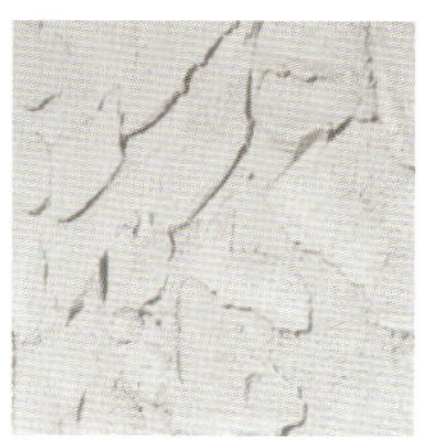
Grey plaster

← View of the front facade

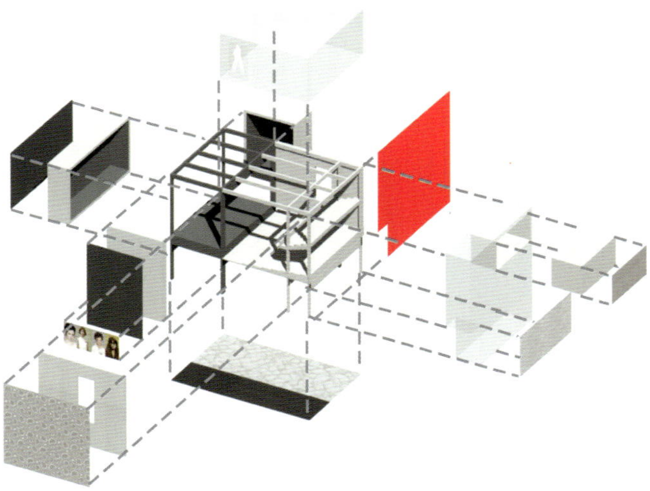

EXPLODED VIEW

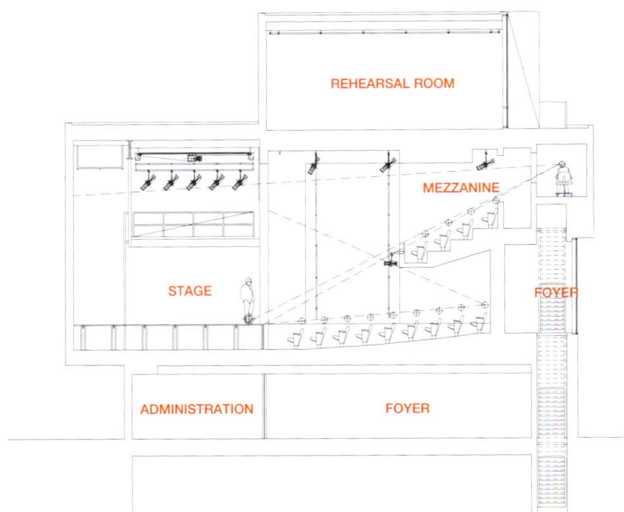

SECTION

1907년 목재와 벽돌로 지어진 허름한 단층 건물의 신축이 불가능하다는 여러 의견에도 불구하고, 방문객들의 안전과 편리를 위해 기존 건물을 철거하고 재건하기로 했다. 극장은 잠깐 머무는 덧없는 공간이지만 함께 했었던 사람들과의 기억과 독특한 순간의 연속으로 우리가 잊을 수 있는 것들의 증거가 되기도 한다. Theatre de Quat'sous 프로젝트를 진행하며 새로운 극장에서 새로운 추억을 환기시키고자 기존 극장의 여러 건축적 장치들을 이용해 달라는 요청을 받았다. 기존의 극장의 질감, 이미지, 색상, 재료를 선택하여 새로운 볼륨(무대, 집, 현관, 교차 공간, 관리 공간, 연습실)으로 구성했다. 기존의 석재, 외장재, 목재, 벽돌, 대리석, 가구들을 재활용 하는 것은 문화적 지속성을 지키는 방법이다. 실크스크린 유리, 검정색 벽돌, 타공 알루미늄 등의 새로운 자재들을 사용하여 추억과 기억이 되살아나는 공간으로 계획했다.

↑ View of the terrace ↙ Facade detail ↘ Facade detail

Coloniale

Exterior view

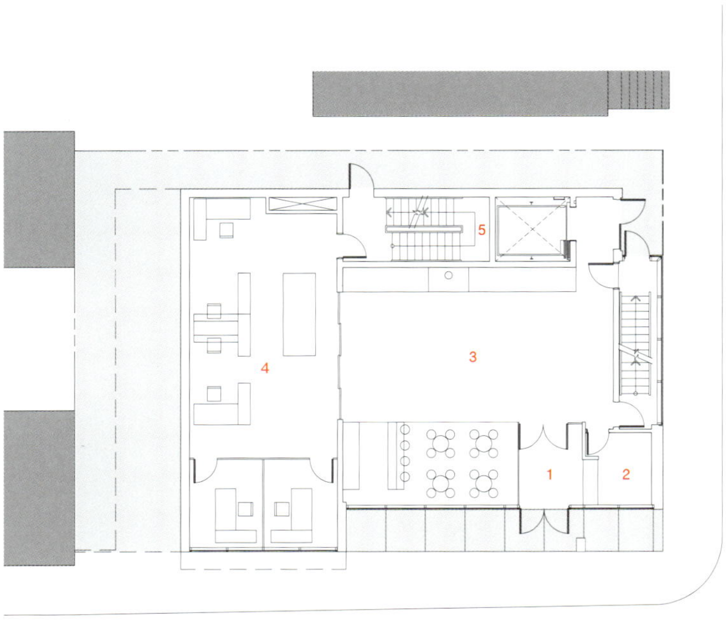

GROUND FLOOR PLAN

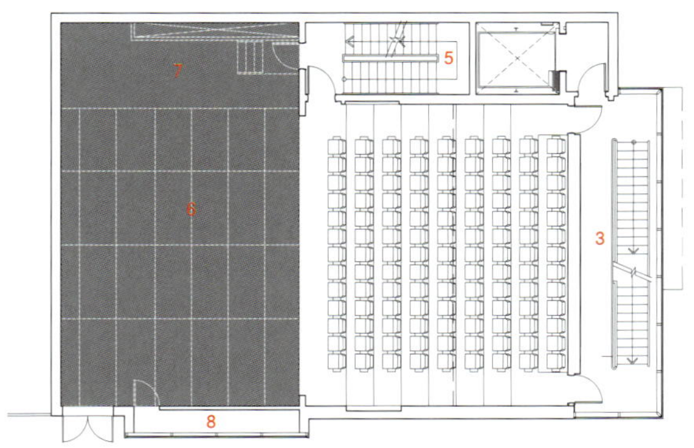

1ST FLOOR PLAN

1 HALLWAY
2 TICKET OFFICE
3 FOYER
4 ADMINISTRATION
5 EMERGENCY STAIRWAY
6 STAGE
7 OFFSTAGE
8 ADVERTISING WINDOW

↰ Theater ↑ Cafe ↓ Administration

← Stairs ↑ Dressing room ↓ Hallway to washrooms and coatroom

AAFE
Community Technology Center
Centro Comunitario de Tecnología

SWITCH BUILDING

스위치 빌딩

Switch Building is a 7 storey apartment and art gallery building at 109 Norfolk Street in the Lower East Side, New York City. The building consists of four floor-through apartments, a duplex penthouse and a double height art gallery on the ground and cellar levels. The "switching" concept opportunistically maximizes difference while maintaining the efficiencies of repetition. In a reinterpretation of a traditional New York bay window, an angled front facade switches back and forth, allowing each floor-through apartment unique views up and down Norfolk Street and creating subtle variations in shadows and reflections. At the rear of each apartment, the living space extends out to large balconies - the largest allowed by zoning, which also shift side to side, creating double-height spaces between balconies to maximize afternoon light and neighborly interactions. While the apartment plans are identical, these variations allow each unit to be unique in its light qualities and views to the city. The new nonprofit Switch Gallery expands its boundaries with the street via a black hot-rolled steel and glass storefront and canopy that open completely to the sidewalk. At the rear of the gallery, visitors descend into a double-height volume illuminated by a large skylight. The gallery's plan maximizes wall space in a fluid spatial continuity, while working around the obstacles of the residential core and lobby with which it shares its footprint.

Architect : nARCHITECTS(Eric Bunge, Mimi Hoang) **Location :** 109 Norfolk Street, New York, USA **Area :** 1,347m²(Art Gallery : 250m² + Apartments : 764m²) **Construction period :** 2004 ~ 2007 **Photographer :** Frank Oudeman, nARCHITECTS

설계 : 엔아키텍츠(에릭 번지, 미미황) **위치 :** 미국 뉴욕 노퍽가 109 **면적 :** 1,347㎡(갤러리 : 250㎡ + 아파트 : 764㎡) **공사기간 :** 2004 ~ 2007 **사진 :** 프랭크 오데만, 엔아키텍츠 제공

스위치 빌딩은 아파트 겸 미술관 건물이다. 4층의 주거 시설과 지층에 두 개 층 높이의 미술관과 꼭대기의 복층 펜트하우스로 구성되었다. "Switching(엇갈림)"이라는 콘셉트는 반복의 효율성을 유지하는 동시에 대비감을 극대화시킨다. 뉴욕의 전통적인 밖으로 내민 창문을 재해석하여 각진 정면의 파사드는 앞뒤로 엇갈리며 층마다 노포크 거리를 내려다 볼 수 있게 하고 그림자와 반사로 절묘한 변화를 만들어 낸다. 각 아파트 후면의 발코니로 주거 생활 공간은 확장되고, 각 발코니와 발코니 사이의 이층 높이의 공간은 오후의 일조량을 확보하며 이웃 간의 소통도 유발시킨다. 각 아파트의 평면은 동일하지만 발코니 구조를 통해 도시를 향한 경관과 채광의 독특함으로 다양성을 부여했다. 지층의 스위치 갤러리는 검은 강철과 유리로 된 매장 정면과 보도 쪽으로 완전히 열린 캐노피를 통해 거리로 그 경계를 넓힌다. 갤러리 뒤쪽에서 방문자들은 커다란 천창이 특징인 이층 높이의 볼륨으로 내려갈 수 있다. 갤러리는 벽을 통한 공간 구성으로 지속적인 유동성을 최대화하며, 많은 사람들의 이동으로 인한 거주 공간과 로비의 장애물을 완화해준다.

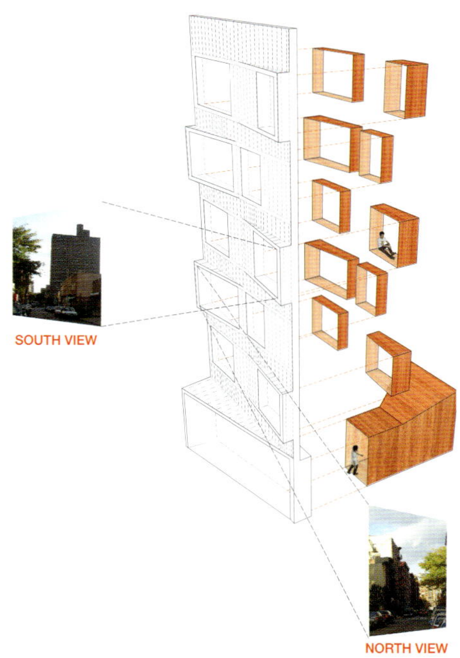

FACADE VIEW DIAGRAM

← Exterior view

↑ Protruding terrace as seen from below ↱ Front facade from below ↳ Entrance

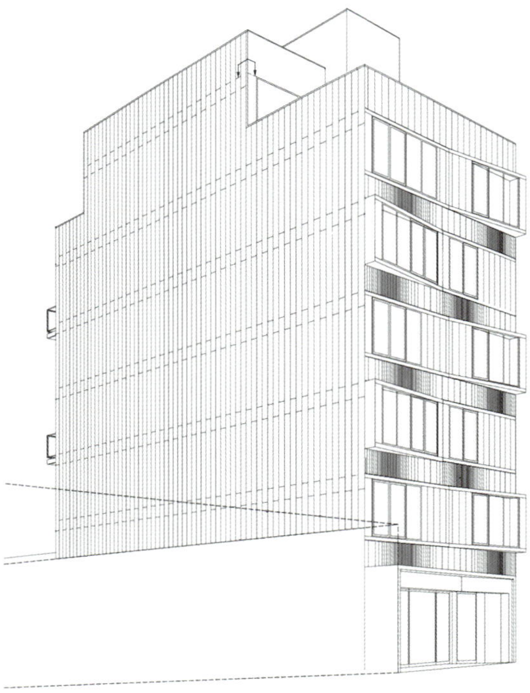

SOUTH EAST VIEW

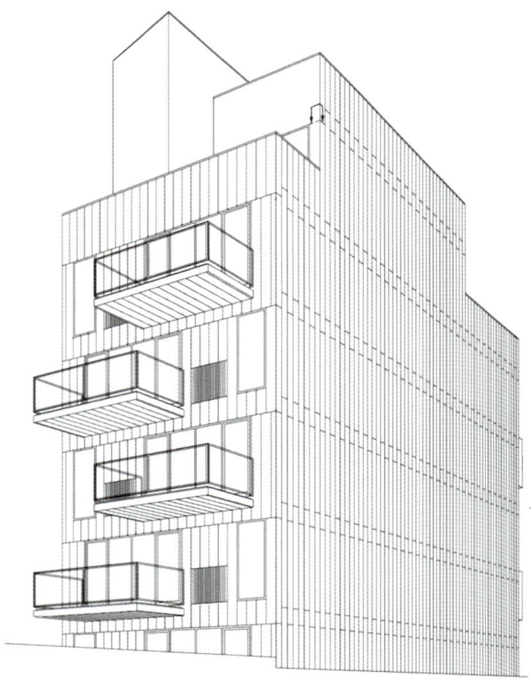

SOUTH WEST VIEW

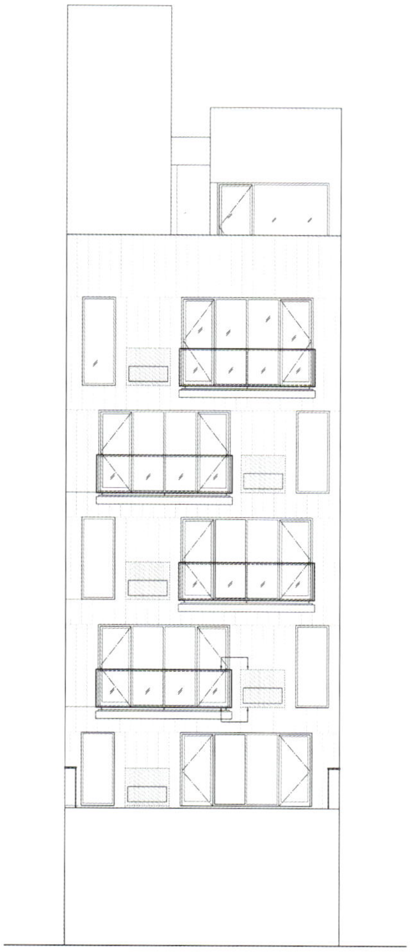

REAR ELEVATION

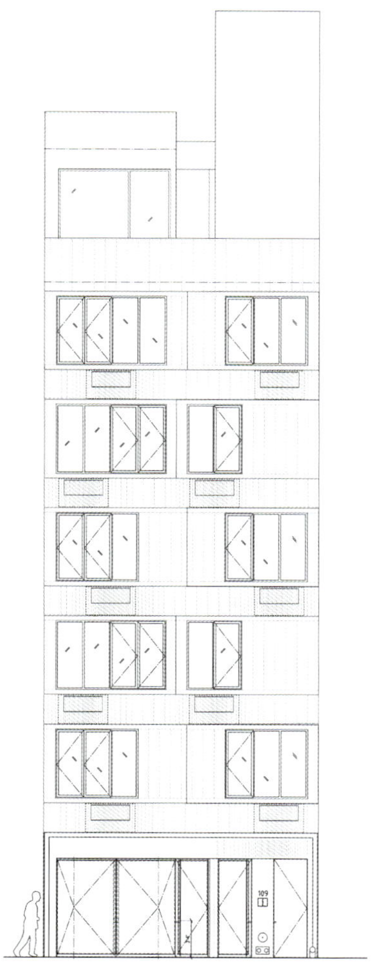

FRONT ELEVATION

← Lobby → Lobby

1 GALLERY
2 LOBBY
3 BEDROOM
4 KITCHEN
5 LIVING ROOM

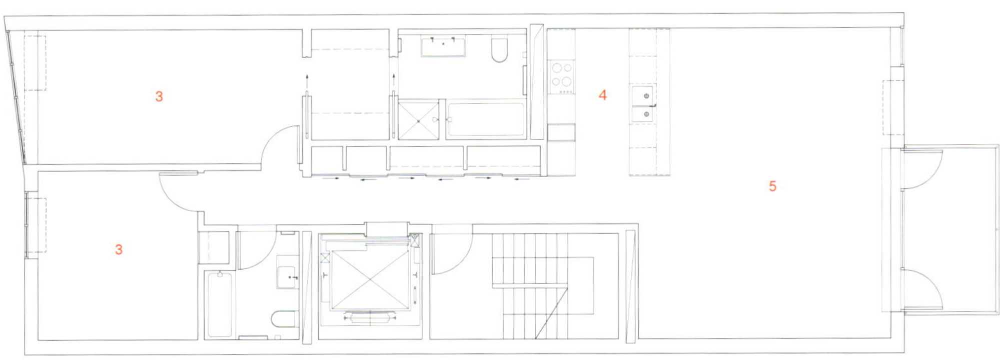

APARTMENT FLOOR PLAN

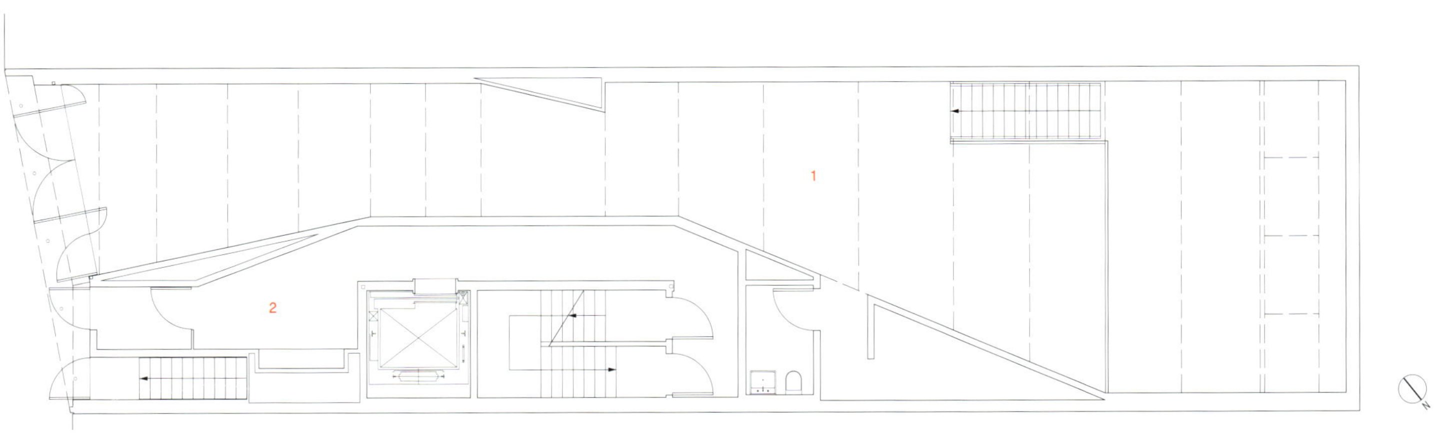

GALLERY FLOOR PLAN

↑ View of the kitchen → View from the stairs

Paragon

M+

엠플러스

M+ is a building located at the resonating border/liminal space where heterogeneous urban elements such as super-luxury residential skyscrapers, dilapidated low-rise housing, houses converted into offices, and offices that double as residences all collide and interact. M+ is a process of discovering the ever renewing relational possibility within this liminality. Like the primal vitality of the colossal black obelisk in the movie 2001:A Space Odyssey, M+ is also a process of unearthing hidden life forces in this urban desert, or like the forest spirits of the animation Princess Mononoke, it is a process of dematerialization as the awakened genius loci with so many tentacles dissolves the mass of urban concrete. Also, M+ strives to restore the relationship between nature and human beings that has been lost in the desolate urban environment. This is done by uncovering and pulling visual corridors for nature into the site. As for the general functional program, I agonized over the 3D puzzle of creating 'a house with garden' within four-story space. This is because the most popular space in the neighborhood is the renovated residential office, namely a house with garden. I saw that a vertical "house with a garden" would provide more flexibility in accommodating various rental needs. To make this 3D spatial organization possible, the usually forgotten (in terms of design) egress stair is peeled off from a core, and renewed as vertical open space with greenery, threading through various voids inside and outside the building. This space includes the sunken garden, second floor courtyard, fourth floor terrace, and roof garden. Such a vertical garden creates new relationships among users as well as between users and the neighborhood, architecture and the city, and ultimately between people and nature. Also the corner of the building above the sunken garden is lifted to allow a new relationship between M+ and the surrounding urban setting at the corner, and even more importantly, between the private and public domains.

Architect : HAHN Design(Hahn Joh) **Location :** 251 Nonhyeon-dong, Gangnam-gu, Seoul, Korea **Site area :** 611.5m² **Building area :** 345.97m² **Gross floor area :** 1,269.63m² **Building coverage :** 56.58% **Floor space index :** 149.87% **Construction period :** 2008. 7 ~ 2009. 8 **Photos offer :** HAHN Design

설계 : 한디자인(조한) **위치 :** 서울특별시 강남구 논현동 251번지 **대지면적 :** 611.5㎡ **건축면적 :** 345.97㎡ **연면적 :** 1,269.63㎡ **건폐율 :** 56.58% **용적률 :** 149.87% **공사기간 :** 2008. 7 ~ 2009. 8 **사진 :** 한디자인 제공

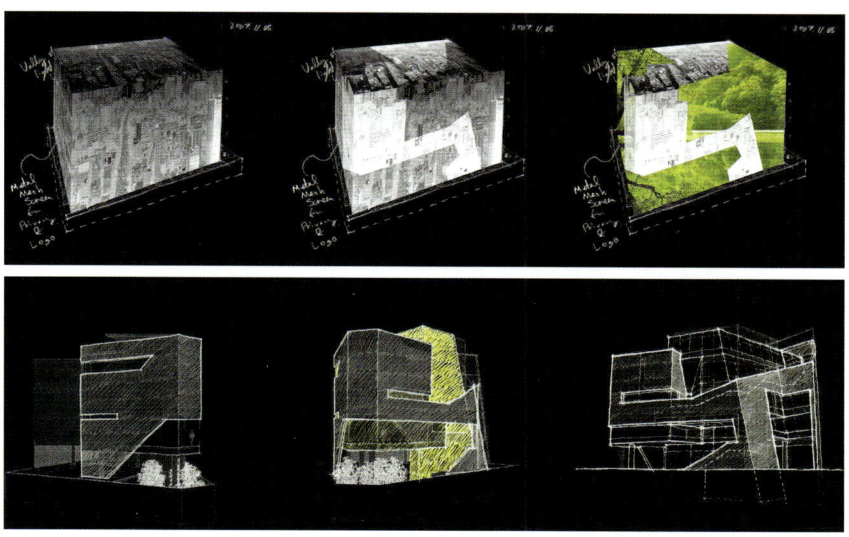

CONCEPT & SKETCH

← Overall view

↑ View from North → View from East

엠플러스는 강남의 거대한 도시 구조 속에서 고층 고급 아파트와 저층 노후 아파트, 오피스화 된 주택과 주거화 된 근린생활시설 등 다양한 이질적 충돌과 관계의 경계에 위치한 건물이다. 이러한 다중적인 경계 속에서 잠재된 새로운 관계의 가능성을 찾는 과정이다. 스탠리 큐브릭 감독의 영화 〈2001 스페이스 오디세이〉에 등장하는 거대한 검은 오벨리스크의 원초적 생명력처럼 삭막한 도시의 사막 속에서 새로운 생명의 가능성을 찾는 과정이며, 만화영화 〈원령공주〉에 나오는 숲의 정령처럼 지하에서 깨어난 땅의 신(Genius loci)이 수많은 촉수를 통해 도시의 콘크리트 볼륨을 해체하는 비물질화의 과정이기도 하다. 엠플러스는 삭막한 도시적 환경 속에 숨겨진 자연과 인간의 관계를 찾기 위해, 밀집한 주변 건물들 사이로 보이는 자연과의 시각적 통로를 대지 내로 끌어들여 자연과 인간의 관계성을 회복하고자 한다. 전체적인 프로그램 구성 면에서 '마당이 있는 주택'과 같은 공간을 4층 공간 내에서 3차원 퍼즐처럼 어떻게 풀 수 있는지 고민했다. 이는 인근에서 제일 임대가 잘되는 개조한 주택형 오피스 즉 '마당이 있는 주택'을 참고한 것인데, 수직으로 구성된 '마당이 있는 주택'이라면 수평과 수직 방향의 다양한 임대와 주거의 방식이 가능해진다. 이러한 3차원적 공간 구성을 실현하기 위해, 일상적으로 디자인 측면에서 버려지곤 했던 근린생활시설 내부의 피난계단을 코어에서 분리하여, 선큰가든, 2층 중정, 4층 테라스, 옥상정원 등 건물 내외부의 다양한 보이드 공간을 연결하는 수직적인 오픈 공간/마당으로 재탄생 시켰다. 3차원 마당 덕분에 사용자와 사용자, 사용자와 주민, 건물과 도시 그리고 궁극적으로 인간과 자연 사이의 다양한 관계성이 가능해진다. 또한 대지 모퉁이의 선큰가든 위에 필로티로 띄워진 볼륨은 근린생활시설 건물과 도시가 공공성 측면에서 새로운 관계를 찾게 한다.

↑ View from Northeast

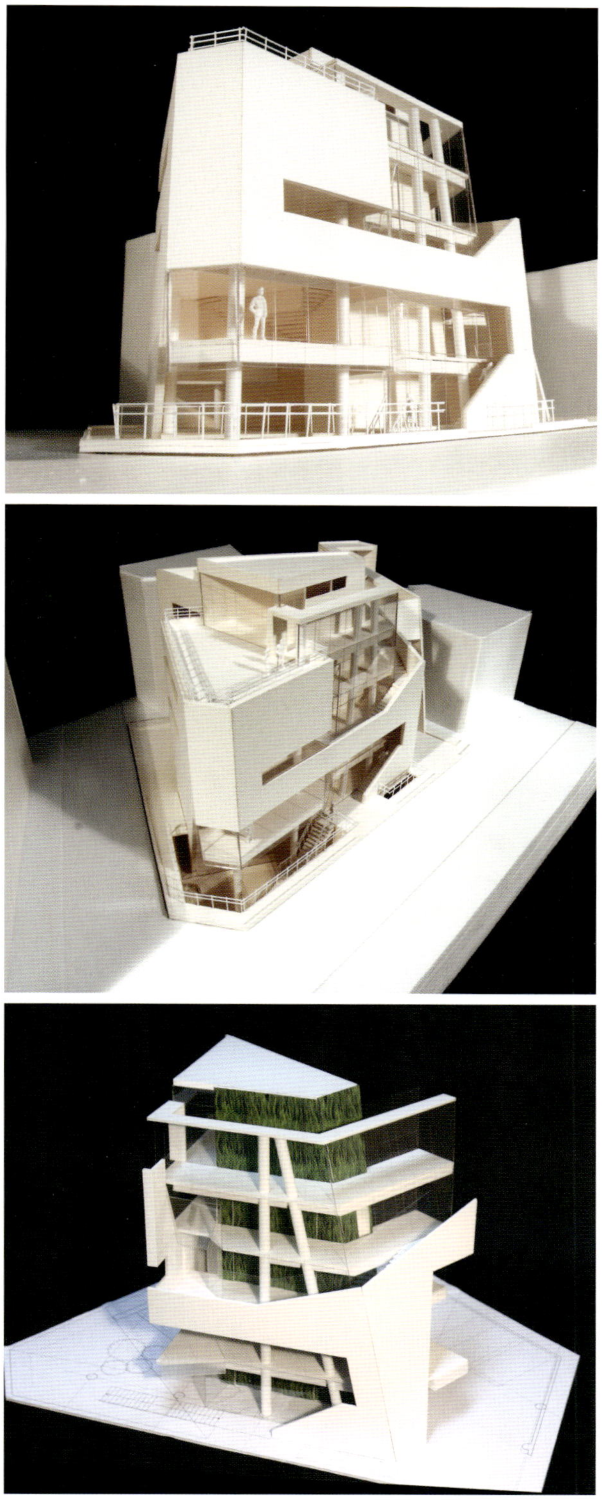

MODEL

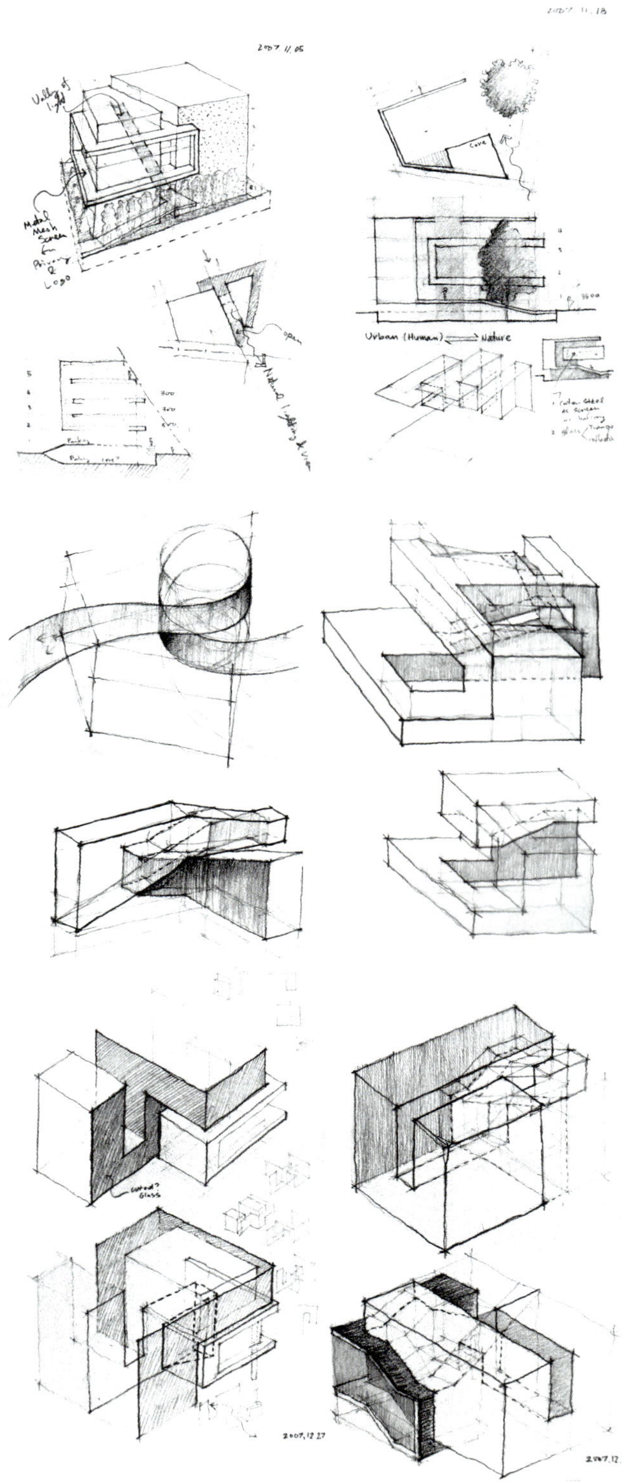

SKETCH

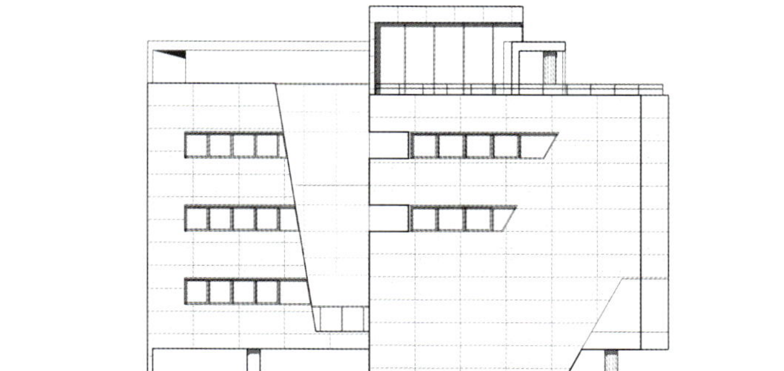

EAST ELEVATION

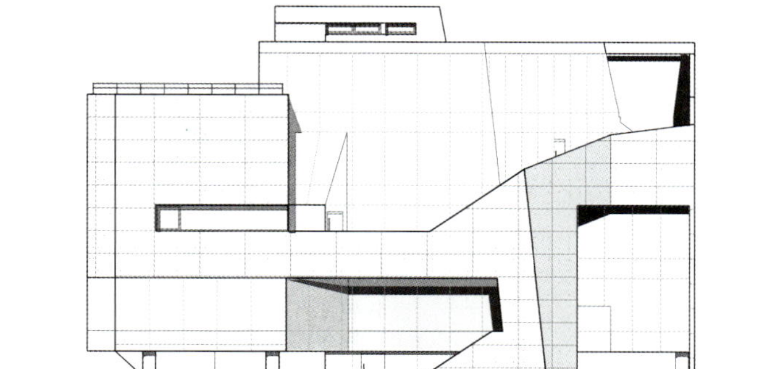

NORTH ELEVATION

← Courtyard ┌ Courtyard └ Connected space

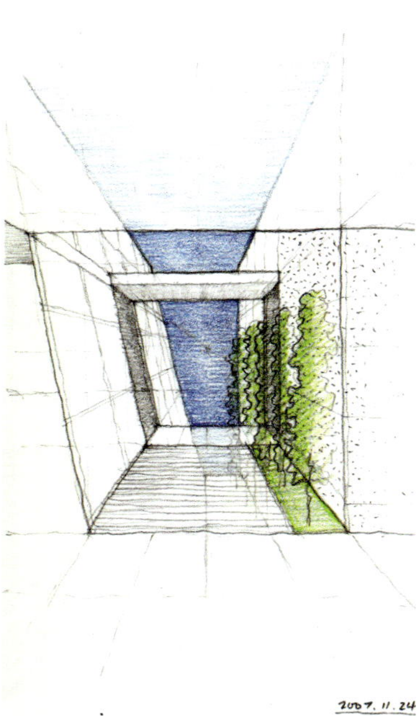

SKETCH

↑ Sunken
↲ Entrance
↳ External stairs on the 3rd floor

↑ Roof garden

1 ROOF GARDEN
2 OFFICE
3 HALL
4 PARKING
5 LOFT
6 CORRIDOR

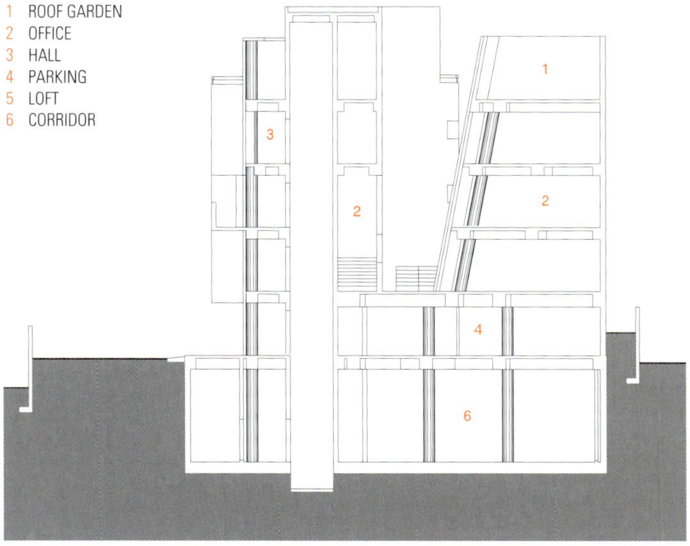

LONGITUDINAL SECTION

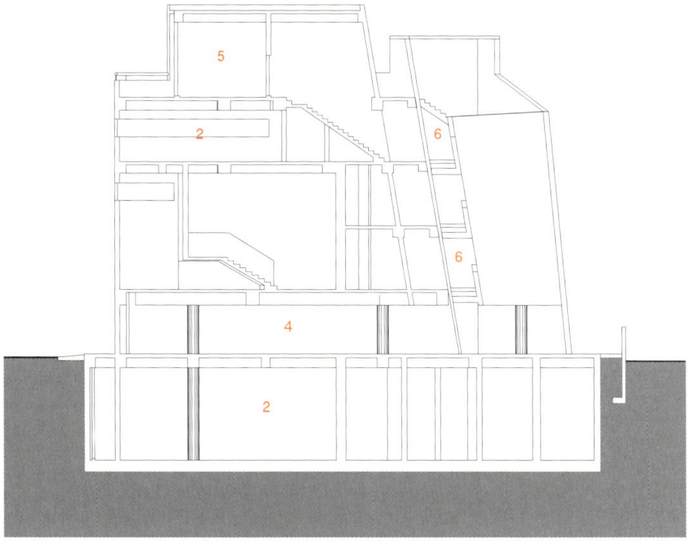

CROSS SECTION

← Corridor of the roof garden ↲ Roof garden

1 PARKING LOT
2 TERRACE
3 HALL
4 MEETING ROOM
5 RETAIL
6 OFFICE

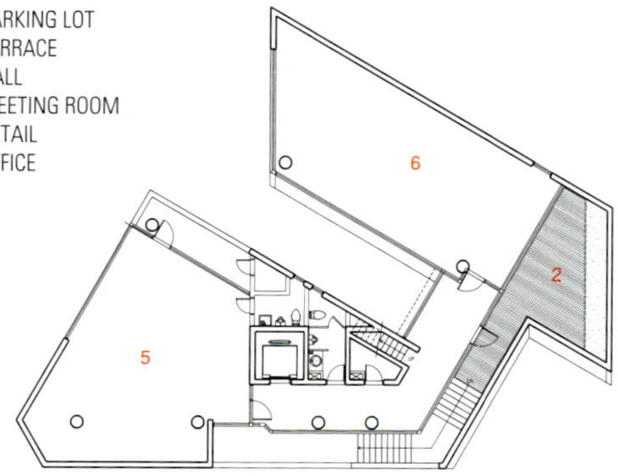

4TH FLOOR PLAN

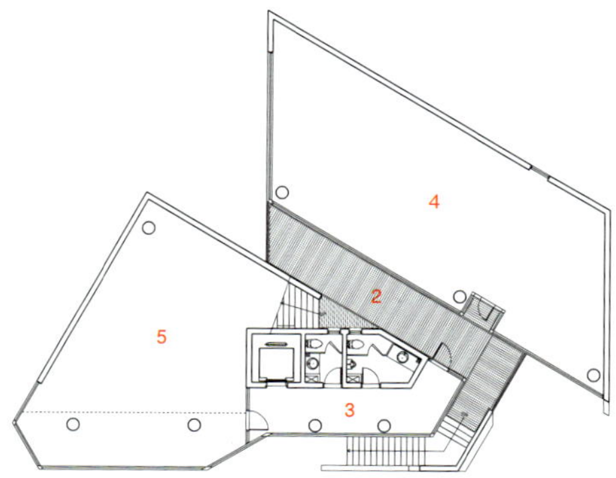

2ND FLOOR PLAN

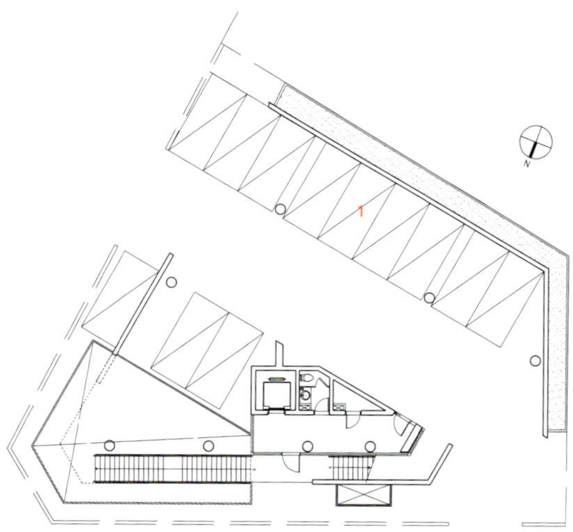

1ST FLOOR PLAN

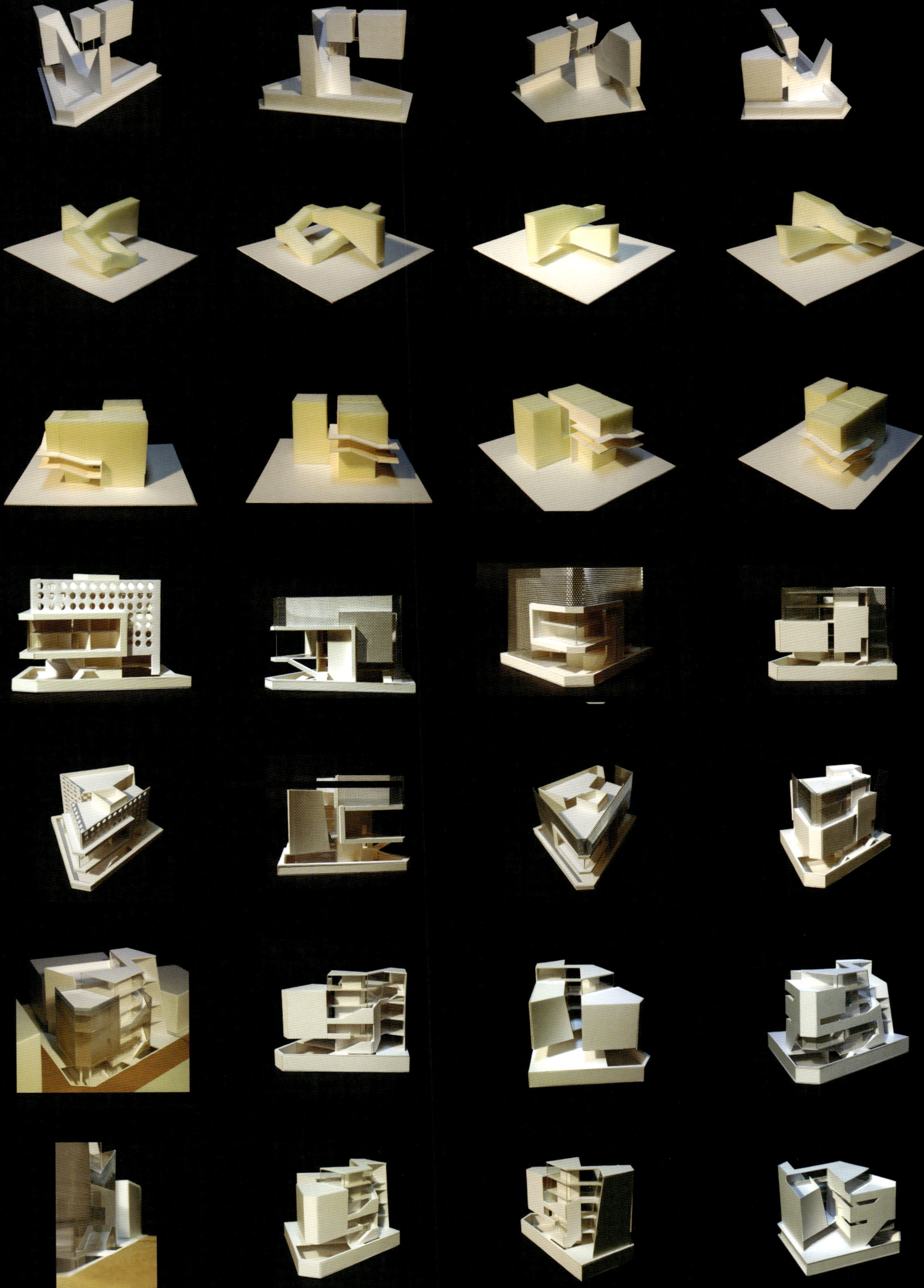

↑ MODEL STUDY

BATI_ㄹ

바티_리을

Modern cities require necessary living facilities to improve the quality of life and to fulfill commercial functions. But the facilities that were built for the residents' convenience ironically snatch living spaces from the residents and finally make them live their homes. Thus, we considered about a small living facility that can activate the community and accept various functions such as commercial function, culture and life pattern. The facility can establish a new relationship between the horizontal spaces and the architecture. The architecture was differentiated from other buildings which are consisted of vertically disconnected levels and have no convenient access from the streets and was designed as an intersection which has flexible flows. Therefore we aimed to facilitate access to the building through the three-dimensional accesses with expanded ground that can communicate with the surroundings and leads public communities. In the architectural volume, a flexible spatial structure using stairways or slope adequately was preferred to mechanical equipments such as an elevator. In addition, we created a volume in which the exterior can interact with the interior by maximizing use of the artificial ground such as sunken, terrace, roof garden and mezzanine. By creating a community in the urban space and connecting the circulation in the building horizontally, we designed the architecture which can transform the entire spatial structure.

Architect : Kim Dong Jin(Hongik University), L'EAU design Co., Ltd. **Location :** 59-12 Cheongdam-dong, Gangnam-gu, Seoul, Korea **Site area :** 274.62m² **Gross floor area :** 742.88m² **Completion :** 2008. 7 **Photo grapher :** Park Wan Soon

설계 : 김동진(홍익대학교), (주)로 디자인 도시환경건축연구소
위치 : 서울특별시 강남구 청담동 59-12 **대지면적** : 274.62㎡
연면적 : 742.88㎡ **완공** : 2008. 7 **사진** : 박완순

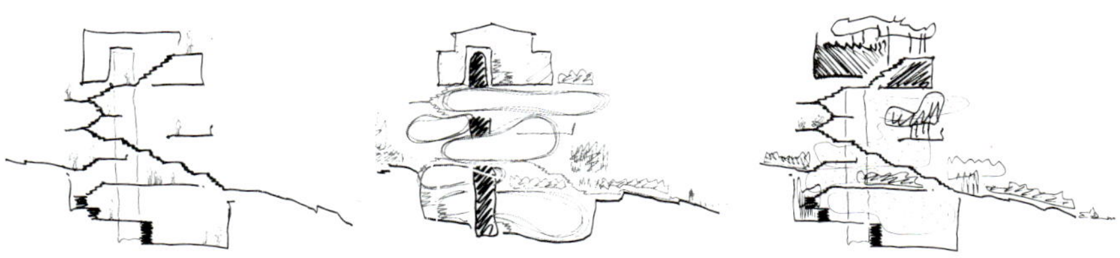

DIAGRAM

← View from Northeast

1 PARKING LOT
2 NEIGHBORHOOD FACILITY
3 ENTRANCE HALL

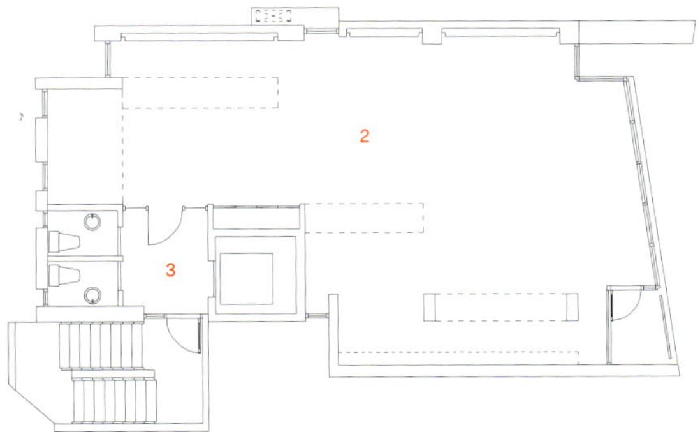

4TH FLOOR PLAN

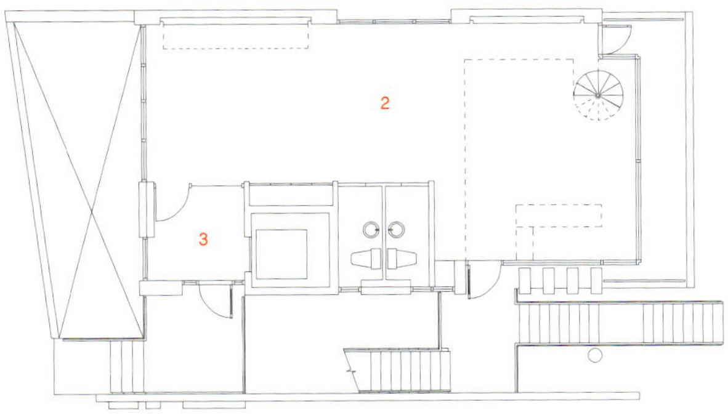

2ND FLOOR PLAN

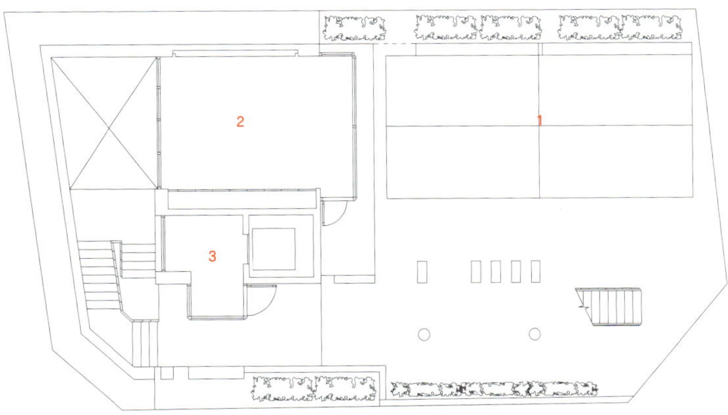

1ST FLOOR PLAN

← View of front facade ↑ View from sloped terrace ↓ External stairs

현대 도시는 사업성을 충족시키는 동시에 도시 내 공공성을 바탕으로 주거민의 생활의 질을 향상시킬 수 있는 근린 생활 시설을 필요로 한다. 시작은 편의를 위해 필요했던 근린 생활 시설이 때로는 주민의 생활 터전을 빼앗고 끝내 이곳을 떠나게 만든다. 따라서 거주 환경의 근린성 보장을 위한 구체적인 대안과 지역 커뮤니티를 활성화와 상업적 기능 및 문화와 생활 패턴을 수용할 수 있는 공공성이 배려된 소규모 근린 생활 시설의 연구는 가로 공간과 건축물 사이의 새로운 관계를 설정할 수 있을 것이다. 도시 내 주거지에서 가로 공간과 상호작용 할 수 있는 가로로부터 접근성의 부재와 수직적 층별 단절된 공간 구조를 탈피하는 유동적 흐름을 가진 교류 가능한 장의 역할을 하도록 계획하였다. 따라서 가로의 연속성을 통해 소통할 수 있고 공공의 커뮤니티를 유발하도록 지면층 확장에 의한 입체적 진입으로 접근성이 증대되도록 하였다. 건축 볼륨 안에서 엘리베이터와 같은 기계적 장치에 의존하기보다 계단 또는 경사로의 적절한 활용으로 건물 전체를 원활하게 움직일 수 있는 공간 구조 체계를 마련하였다. 또한 선큰, 테라스, 옥상 정원, 메자닌 등의 인공적 지면 활용의 극대화로 내외부 공간 안에서 상호 작용이 일어나는 볼륨을 가지도록 하였다. 도시 공간 내 커뮤니티를 만들고 가로로부터 건물 내부 동선 체계까지 상호 작용하여 건축물의 전체적인 공간 구조 변화의 가능성을 가진 건축물을 계획하였다.

↑ The part of view from East ↱ The part of view of the facade ↳ The part of view of the facade

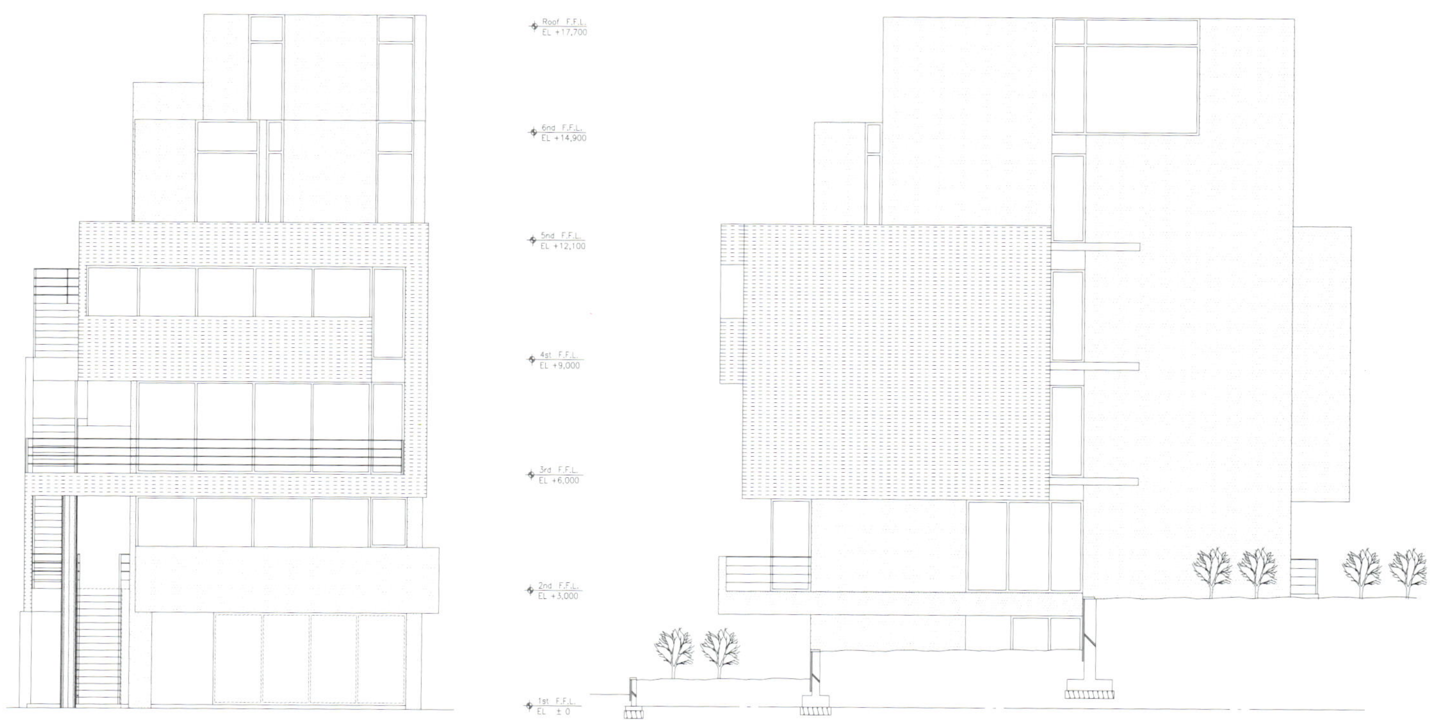

NORTH ELEVATION

EAST ELEVATION

SKETCH

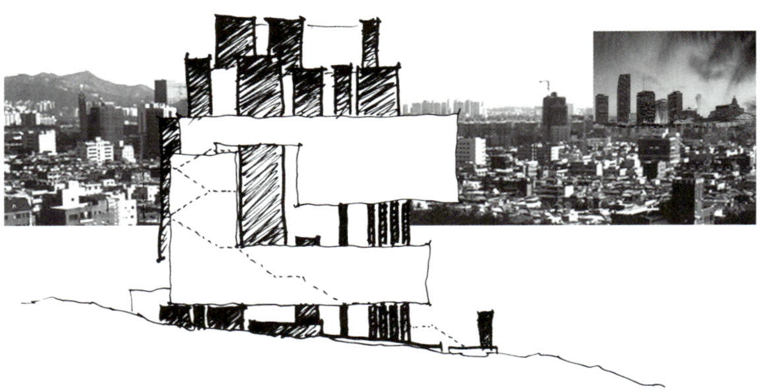

DIAGRAM

SKETCH

⌝ View from Southeast
⌜ Stairs below space
⌞ Stairs space

↑ Interior view
↲ Stairs space
↳ Stairs space
↱ View of the 1st floor

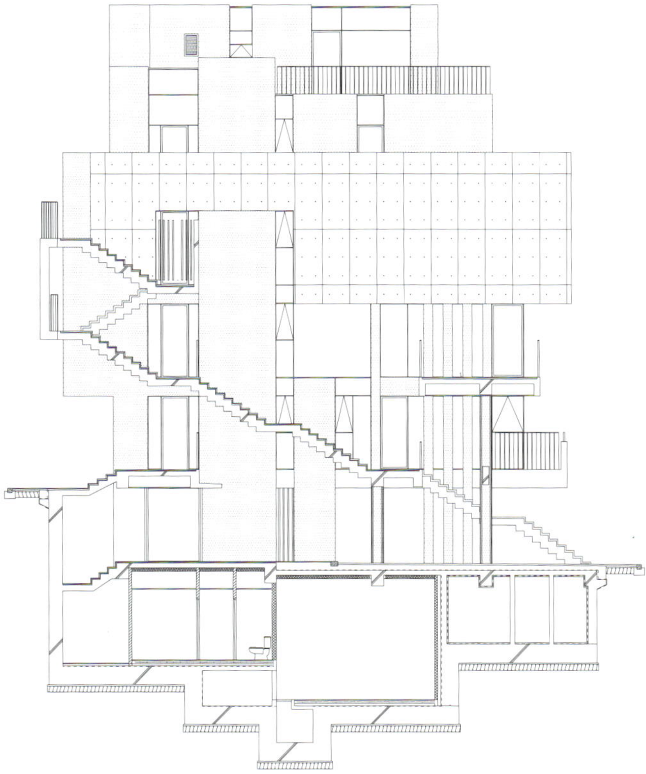

SECTION

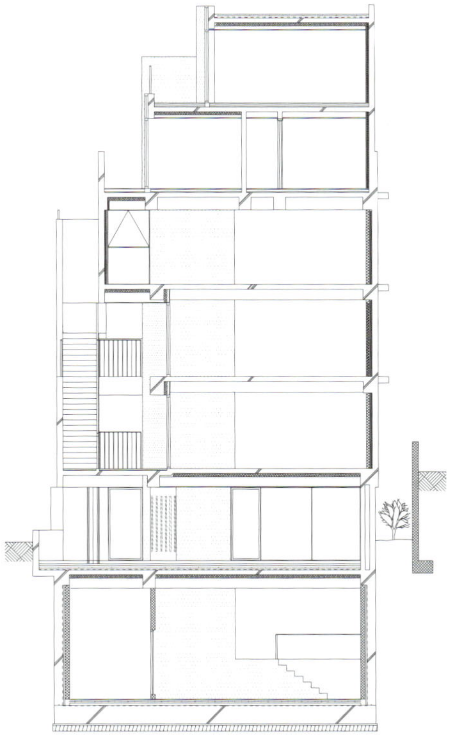

SECTION

BYGAMI
바이가미
DIAMOND
WEDDING RING
COUPLE RING
BYGAMI
gia
party & flowers
544-7044

CHEONGDAM-DONG 81-5 BUILDING

청담동 81-5빌딩

On the hill in Cheongdam-dong where an old house had been demolished, a new building was built. The client, a well-known fashion designer, wanted the lower part of the building to be a rental space with a studio and the upper part to be a residential house with his working studio. It was important to use the not-so-big land well and a large workplace with views over the hill and a residential space were required. Commercial facilities were arranged in the sunken on the basement and stairs to access each floor directly were arranged outside of the building. Stairs go up to the upper house. The building looks like a cube with a roof on top. The roof was planned to maximize spatial efficiency reflecting complex street width regulation. Several roof pieces that surround the rooftop garden for small parties are overlapped with each other and outside views are seen through small openings between them. Covered parts protect the family's privacy. There are balconies on each floor where residents can plant trees. The overall volume was finished with exposed concrete and basalt and the roof was made of panels with similar textures to those two materials.

청담동 언덕 위에 오래된 주택이 헐리고 새 건물이 들어섰다. 잘 알려진 패션 디자이너인 건축주는 저층부를 스튜디오를 겸한 임대공간으로, 상층부는 작업 스튜디오를 겸한 주택으로 만들기를 원했다. 크지 않은 대지를 활용하는 것이 중요했고, 언덕 위 전망을 잘 즐길 수 있는 넓은 작업 공간과 주택 공간이 필요했다. 지하 선큰에 상업시설을 배치했고, 각 층마다 직접 접근이 가능하도록 계단을 건물 밖으로 배치했다. 계단은 상부 주택으로 조용히 이어진다. 건물은 반듯한 육면체 위에 지붕이 얹힌 모습이다. 복잡한 사선 제한을 그대로 반영하여 공간 효율을 최대로 활용하도록 지붕을 계획했다. 작은 파티가 열릴 옥상정원을 감싸는 몇 개의 지붕 조각들은 서로 겹쳐지고 그 사이로 주변의 풍경은 변화감 있게 보인다. 조금씩 가려진 부분으로 프라이버시는 확보된다. 각 층마다 나무를 심을 수 있는 발코니를 계획했다. 노출콘크리트와 현무암으로 전체 볼륨을 구성했으며, 지붕은 이 두 재료와 같은 느낌의 패널로 만들었다.

Architect : Kwon Moon Sung(Sungkyunkwan University) + ATRELIER 17 **Location :** 81-5 Cheongdam-dong, Gangnam-gu, Seoul, Korea **Site area :** 369.60m² **Building area :** 219.86m² **Gross floor area :** 892.21m² **Building coverage :** 59.51% **Floor space index :** 178.71% **Construction period :** 2007. 10 ~ 2009. 2 **Photographer :** Mun Jeong Sik

설계 : 권문성(성균관대학교) + (주)아뜰리에17 **위치 :** 서울 특별시 강남구 청담동 81-5 **대지면적 :** 369.60㎡ **건축면적 :** 219.86㎡ **연면적 :** 892.21㎡ **건폐율 :** 59.51% **용적률 :** 178.71% **공사기간 :** 2007. 10 ~ 2009. 2 **사진 :** 문정식

← View of the front left side

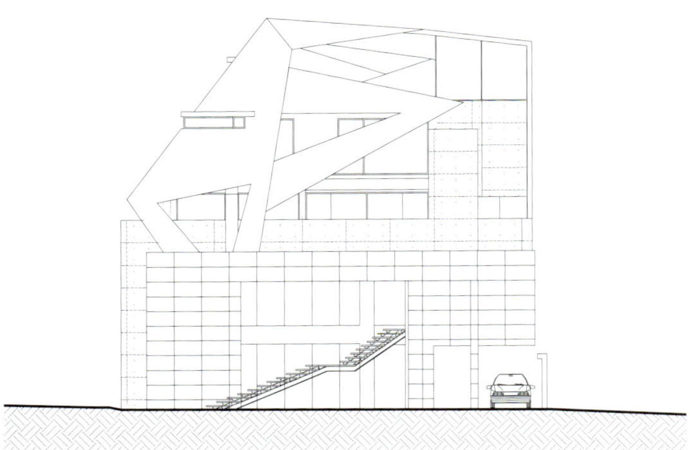

FRONT ELEVATION

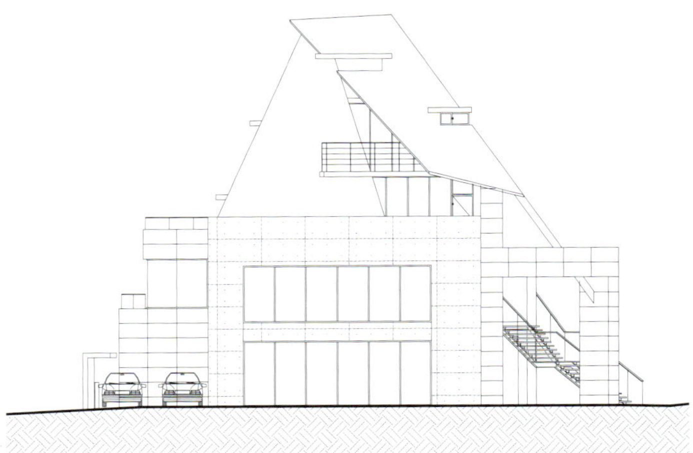

LEFT ELEVATION

General view

Roof garden

View from the roof garden

View from the right side

← External stairs of basement ↑ Roof garden

1 LIVING ROOM
2 BEDROOM
3 NEIGHBORHOOD FACILITY

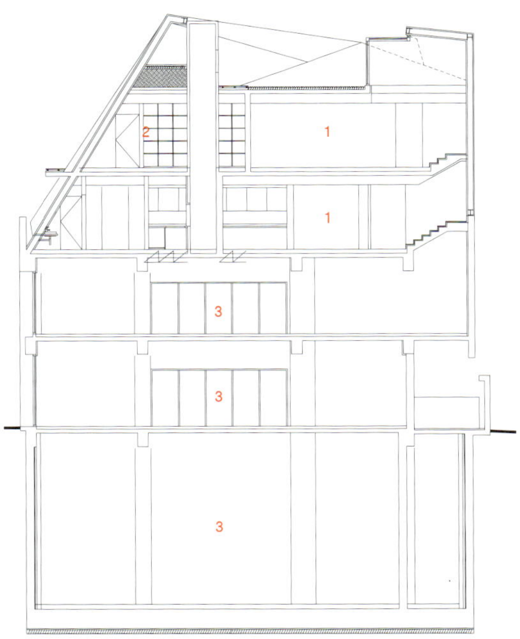

CROSS SECTION

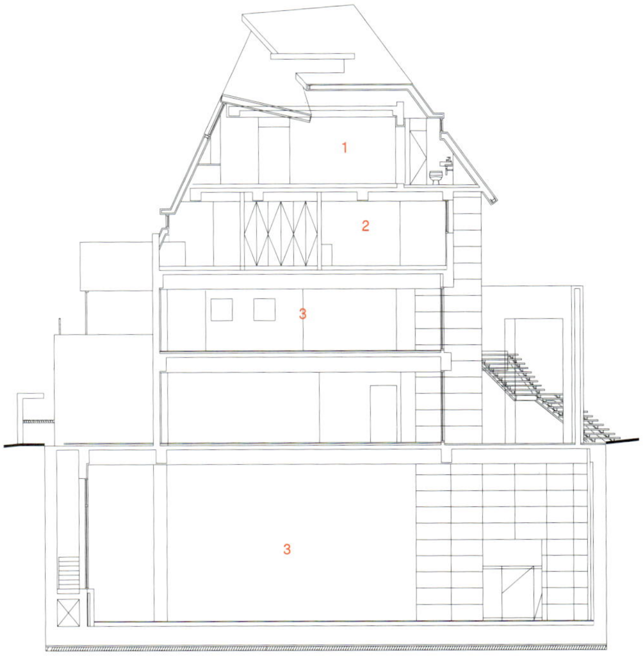

LONGITUDINAL SECTION

← Void space viewed from basement ↰ External stairs of basement ↱ Terrace on the 3rd floor

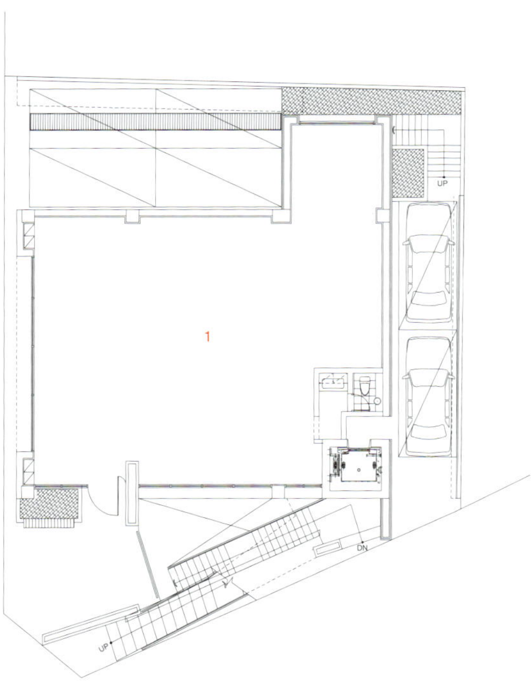

1ST FLOOR PLAN

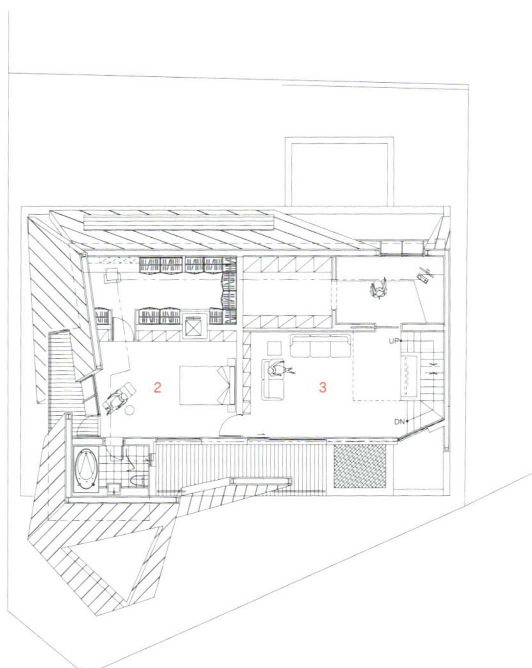

4TH FLOOR PLAN

1 NEIGHBORHOOD FACILITY
2 BEDROOM
3 LIVING ROOM

↰ Living room on the 4th floor ↑ Living room on the 3rd floor ↓ Living room on the 3rd floor

THE STRATA

신사동 637-11

This Strata on a corner in Apgujung-dong, a center of fashion, suggests a new formation among numerous rectangular box-shaped buildings. This building induced three-dimensional curves and its appearance although relatively simple is still impressive. 19 streamlined curves with 54 different radiuses were connected all together and piled to make a dynamic layered space. Two-dimensional streamlined curves make for a three-dimensional flow at elevation and express unique formative aesthetics. In this building, functional distinction of plane spaces is obvious. The public space and the living room are clearly separated by an elevator hall. The elevator hall in a bridge form contains spaces for its front and rear, making the building look bigger. The courtyard at the front entrance is a unique space, which is difficult to access from downtown. It is a transfer space that is connected to the 6th floor between the main entrance and outer spaces to provide a fresh atmosphere to visitors. This overlapped space draws the external streamlined form in and holds an elevation factor soaring up.

Architect : Choi Arcthiects(Charles choi) **Location :** 637-11 Sinsa-dong, Gangnam-gu, Seoul, Korea **Site area :** 378.4m² **Building area :** 184.82m² **Gross floor area :** 1,815.31m² **Design period :** 2006. 2 ~ 2007. 11 **Photographer :** Namgoong Sun

설계 : (주)초이건축사사무소(최철수) **위치 :** 서울특별시 강남구 신사동 637-11 **대지면적 :** 378.4m² **건축면적 :** 184.82m² **연면적 :** 1,815.31m² **설계기간 :** 2006. 2 ~ 2007. 11 **사진** : 남궁선

신사동 637-11은 무수한 직사각형 박스 건물 사이에서 새로운 형태를 제시한다. 3차원 곡선을 도입한 건물의 외관은 단순하면서도 인상적이다. 54개의 반지름을 가지는 유선형의 19개의 곡선들은 서로 연결되고 층층이 쌓여 역동적인 공간을 만든다. 2차원의 유선형 곡선들은 입면에서 3차원적인 하나의 흐름을 만들고, 독특한 조형미를 표현한다. 건물은 평면 공간의 기능적인 분리가 뚜렷하다. 공용 부분과 거실 부분은 엘리베이터 홀을 통해서 확실히 구분된다. 브리지 형태로 연결된 엘리베이터 홀은 앞뒤로 공간을 내포하여 건물 자체가 커 보이는 역할도 겸한다. 출입구 앞의 중정은 도심에서 접하기 어려운 독창적인 공간이다. 주출입구와 외부 공간 사이에 6층까지 연결된 전이 공간으로 방문객에게 새로운 느낌을 경험하게 한다. 이 중첩된 공간은 외부의 유선형 형태가 내부까지 도입되어, 휘몰아쳐 올라가는 입면의 구성 요소도 함께 가진다.

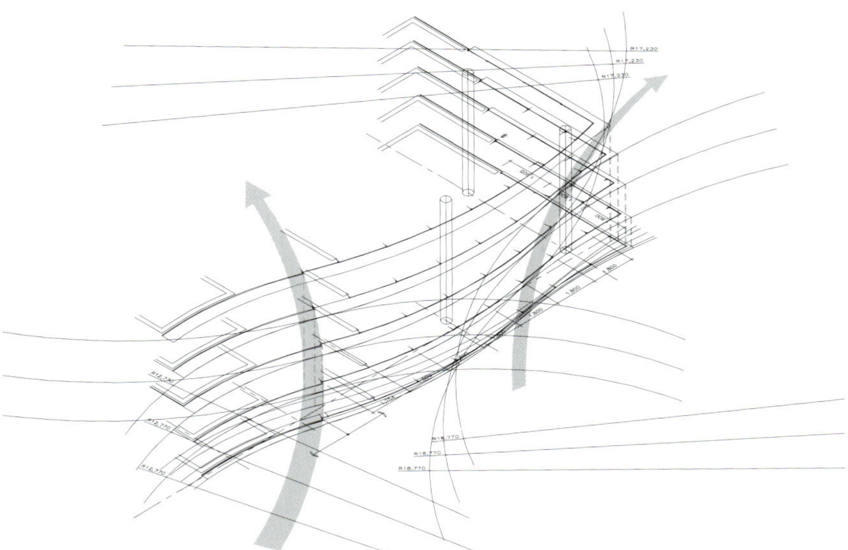

CURTAIN WALL AXONOMETRIC

← Exterior view

↑ View of front facade

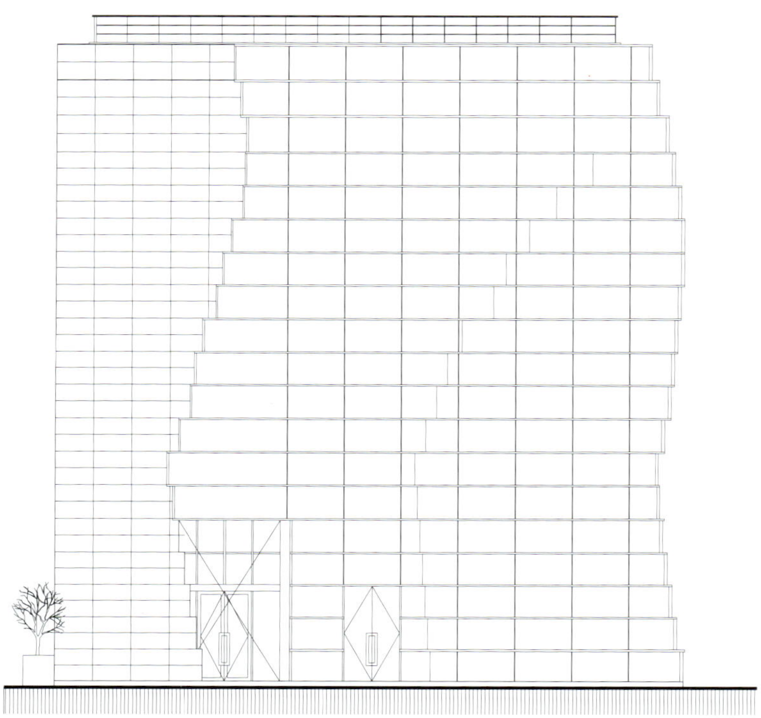

FRONT ELEVATION

RIGHT ELEVATION

↑ Facade detail from the below ↵ A view of the surrounding site ↳ A view of the surrounding site

← View of the elevator hall ↰ Facade view from inside ↱ Facade detail

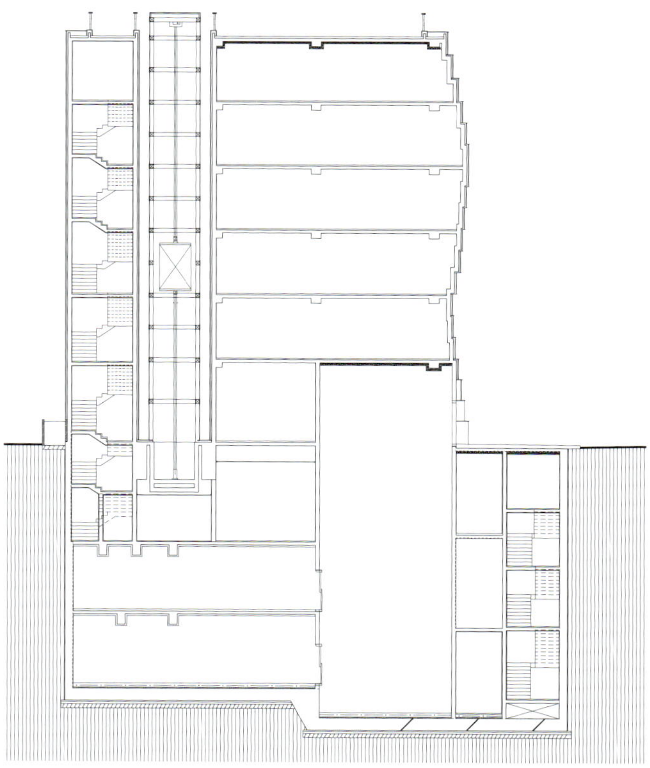

SECTION

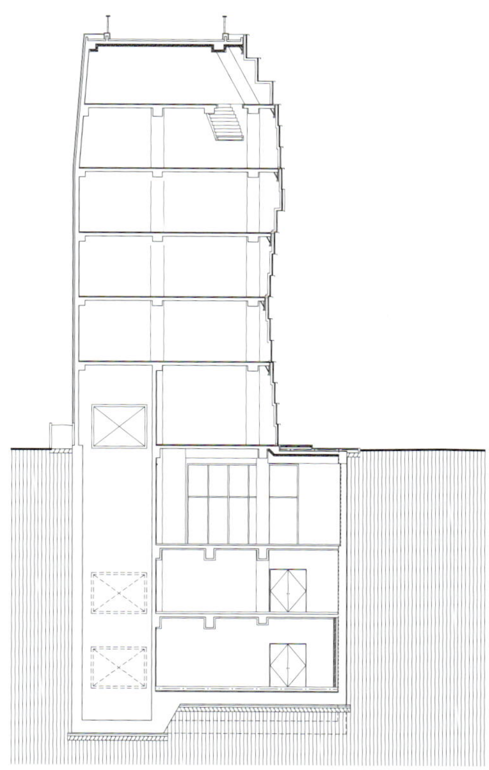

SECTION

← Top-down view of the elevator hall ↑ Interior view from below

1 RESTAURANT
2 ELEVATOR HALL
3 OFFICE

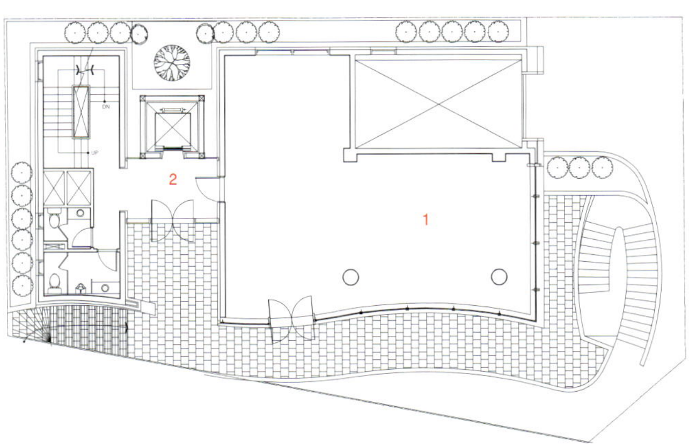

1ST FLOOR PLAN

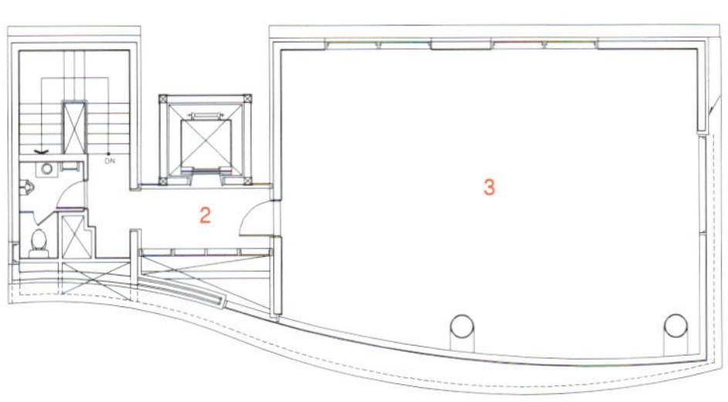

6TH FLOOR PLAN

THE ARC BUILDING

신사동 640

The building's distinctive geometry is not only due to aesthetic concerns but also due to the need to maximize floor space within the allowable setback lines. It sits on a small site, but it contains building volume six times the area of the site since it is located in one of the busiest and trendiest neighborhood of Apgujeong, where the land price is extremely high. Within the given building volume, there are many different functions - a mechanical car park, a gallery at basement levels, a coffee shop on first two levels, and offices from third to sixth levels. This type of mixed-use, medium size commercial building is not usual in Korea where land is scarce and urbanization has progressed rapidly. The challenge in this project has been firstly, to produce a highly adaptive building with multiple usage ; and secondly, to provide easy access and circulation to different levels ; and thirdly, to make a strong physical presence within generic urban landscape of Seoul.

신사동 640은 심미적인 고려와 함께 상업적인 건물로써 면적의 최대화를 만족시키는 계획 목표를 통해 이와 같은 독특한 형태를 갖게 되었다. 작은 대지에 건축면적의 6배에 달하는 볼륨을 형성하며 유행에 민감한 거리의 특성을 선도하는 디자인을 담아냈다. 과감한 곡선을 도입한 건물의 형태는 주변의 시선을 건물 상층부로 끌어 올리는 동시에 각층마다 유기적이고 역동적인 공간을 연출한다. 1, 2층에는 카페, 3층에서 6층은 사무실, 지하에는 주차장과 갤러리 등의 많은 프로그램을 갖는다. 급속도로 도시화 되고 작은 국토의 한국에서 이런 다목적 시설을 갖은 소규모의 상업 건물은 흔하지 않다. 이 프로젝트의 첫 번째 과제는 다목적 기능에 부합하는 것이며 두 번째는 다른 층과의 원활한 접근을 유도하는 것이며, 세 번째는 서울 도시 경관에 강한 인상을 부여하는 것이다.

Architect : Choi Architects(Charles Choi) **Location :** 640 Sinsa-dong, Gangnam-gu, Seoul, Korea **Site area :** 242.7m² **Building area :** 145.15m² **Gross floor area :** 982.03m² **Design period :** 2005. 6 ~ 2008. 9 **Photographer :** Namgoong Sun

설계 : (주)초이건축사사무소(최철수) **위치 :** 서울특별시 강남구 신사동 640 **대지면적 :** 242.7㎡ **건축면적 :** 145.15㎡ **연면적 :** 982.03㎡ **설계기간 :** 2005. 6 ~ 2008. 9 **사진 :** 남궁선

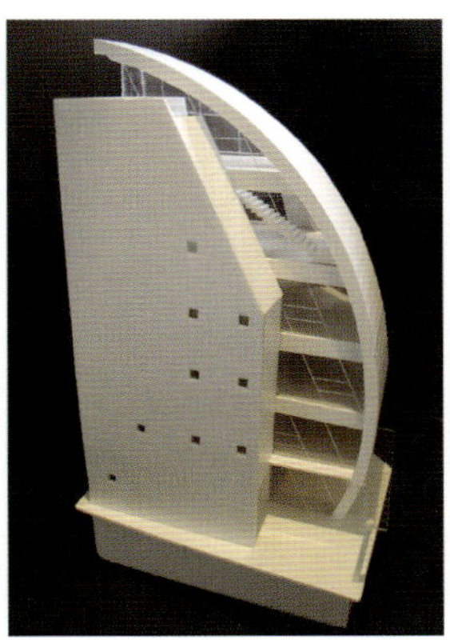

MODEL IMAGE

← View from North

↓ view from Southeast

1 CAR ENTRANCE
2 PARKING LOT
3 SHOP
4 MAIN ENTRANCE
5 NEIGHBORHOOD FACILITY
6 VOID
7 STORAGE

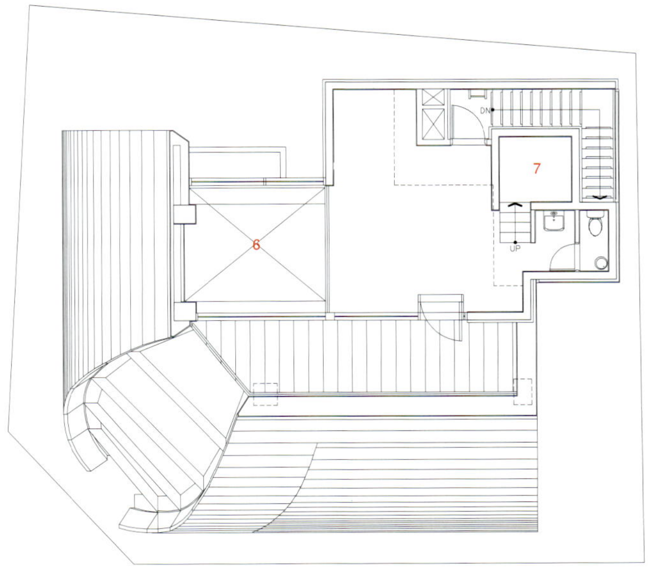

6TH FLOOR PLAN

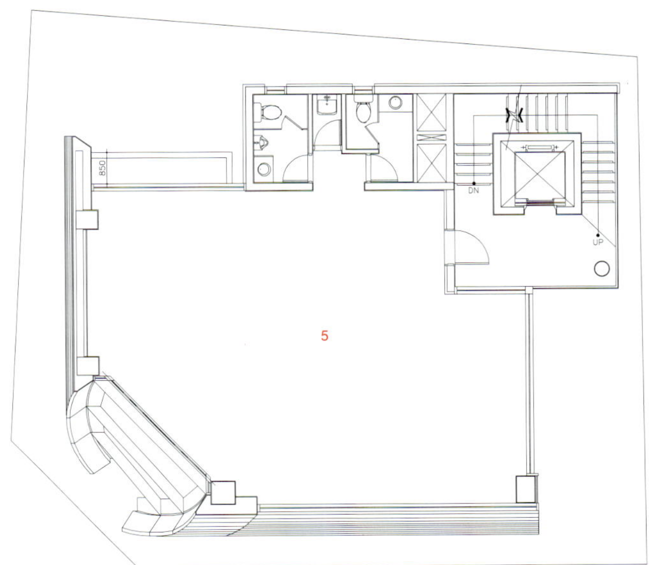

4TH FLOOR PLAN

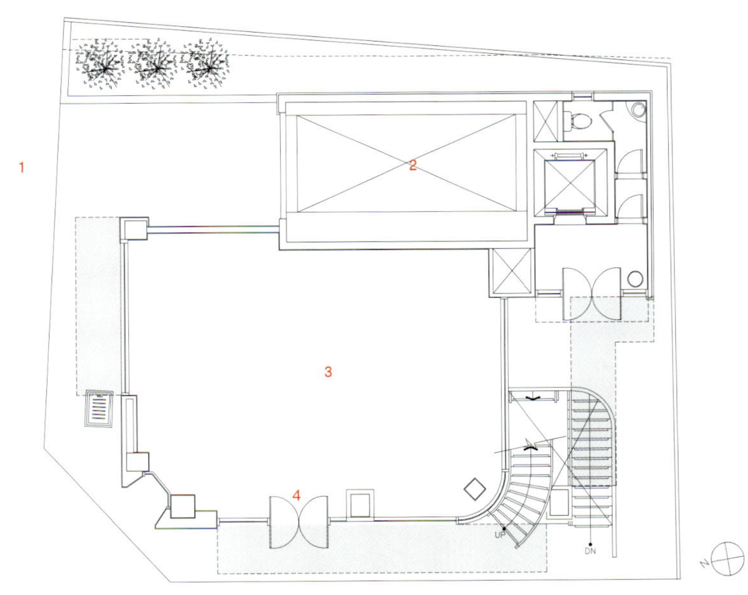

1ST FLOOR PLAN

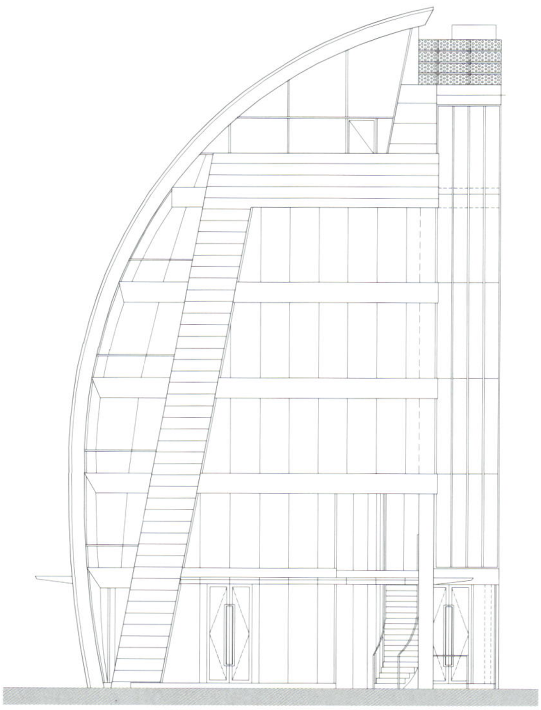

FRONT ELEVATION

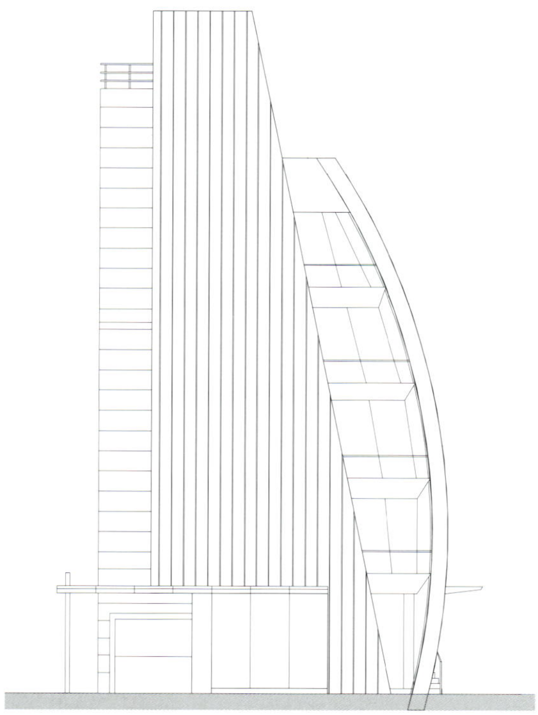

LEFT ELEVATION

← The complete view ┌→ Deck on the 6th floor └→ Entrance area

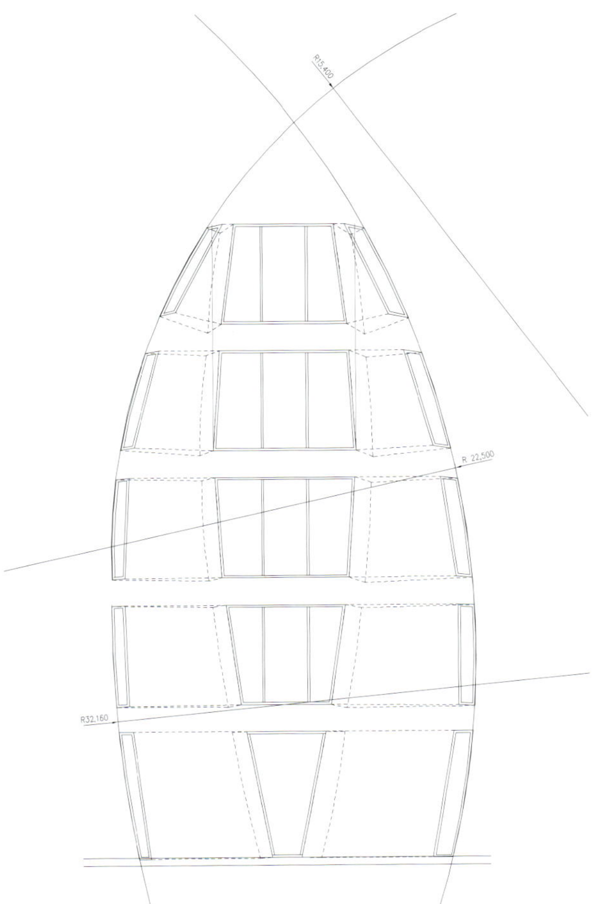

WINDOW DEVELOPMENT FIGURE

CHEONGDAM OFFICE

청담동 125-20

This triangular site facing a narrow street within a dense, non - descriptive neighborhood offered unexpected opportunities to develop a unique structure. Although it contains rather routine program of rentable office and some retail components at ground and basement levels, our goal was to give the building a striking but also welcoming appearance. In order to make the building look more accessible and open towards people passing by, the stair hall is open at the ground level and basement. This is possible by taking out the stair hall walls and supporting the steps around a central column. At higher levels, the stair hall is wrapped in solid, concrete wall. The skin of the building is wrapped in gently curving, ceramic louvers which create dynamic rhythm and usual texture on the facade. It also shields undesirable views from outside at office levels. At the top floor, a duplex is provided with a terrace.

청담동 125-20은 좁은 길과 조밀한 주변 여건을 가진 삼각형의 대지에 위치한다. 이러한 대지의 조건은 독특한 구조를 형성할 수 있는 예상치 못한 기회를 제공하였다. 지하층과 지상층 모두 판에 박힌 소매점들로 구성되지만, 눈에 띄고 매력적인 외형을 형성하고자 했다. 보행자의 접근이 용이하고, 열린 공간으로 보이기 위해 계단실을 외부에 접하도록 계획했다. 건물의 표면을 감싸고 있는 세라믹 루버는 목재의 질감을 표현하면서 역동적이고, 부드러운 곡면을 형성한다. 또한 루버는 사무실에서 낙후된 건물들이 보이는 원치 않는 외부 전경을 가려준다. 꼭대기 층에는 테라스와 함께 복층의 공간을 계획하였다.

Architect : Choi Architects(Charles Choi) **Location :** 125-20 Cheongdam-dong, Gangnam-gu, Seoul, Korea **Site area :** 352.4m² **Building area :** 176.1m² **Gross floor area :** 1,354.78m² **Design period :** 2008. 5 ~ 2009. 1 **Photographer :** Namgoong Sun

설계 : (주)초이건축사사무소(최철수) **위치 :** 서울시 강남구 청담동 125-20 **대지면적 :** 352.4m² **건축면적 :** 176.1m² **연면적 :** 1,354.78m² **설계기간 :** 2008. 5 ~ 2009. 1 **사진 :** 남궁선

← The complete view

woongjin
PC방

← View from Southwest ↰ Night view ↱ Facade view from inside

1 NEIGHBORHOOD FACILITY
2 OUTDOOR PARKING

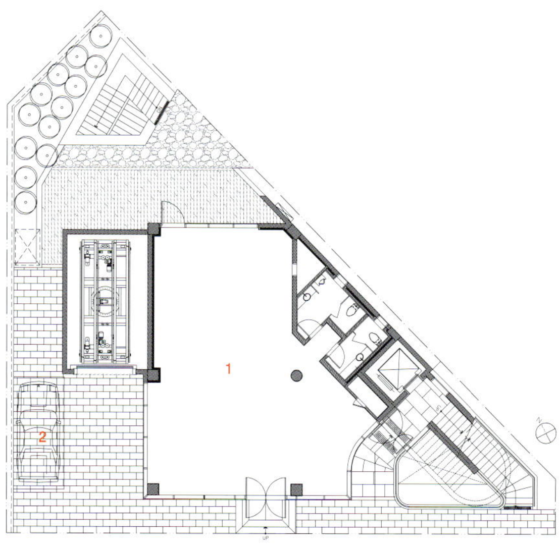

1ST FLOOR PLAN

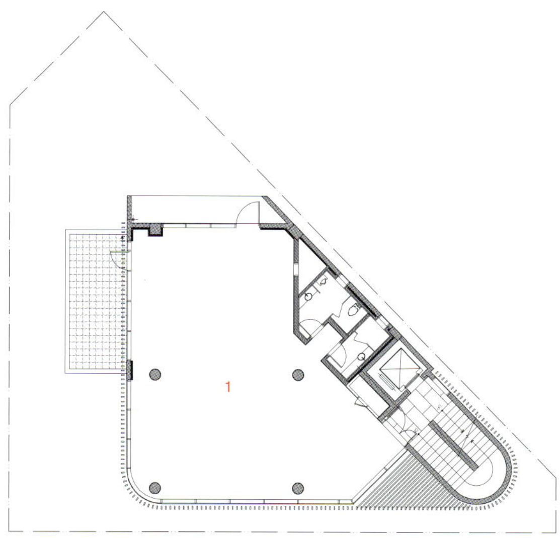

2ND FLOOR PLAN

↑ Overall view showing the urban context ↗ Detail of facade from below ↘ Entrance

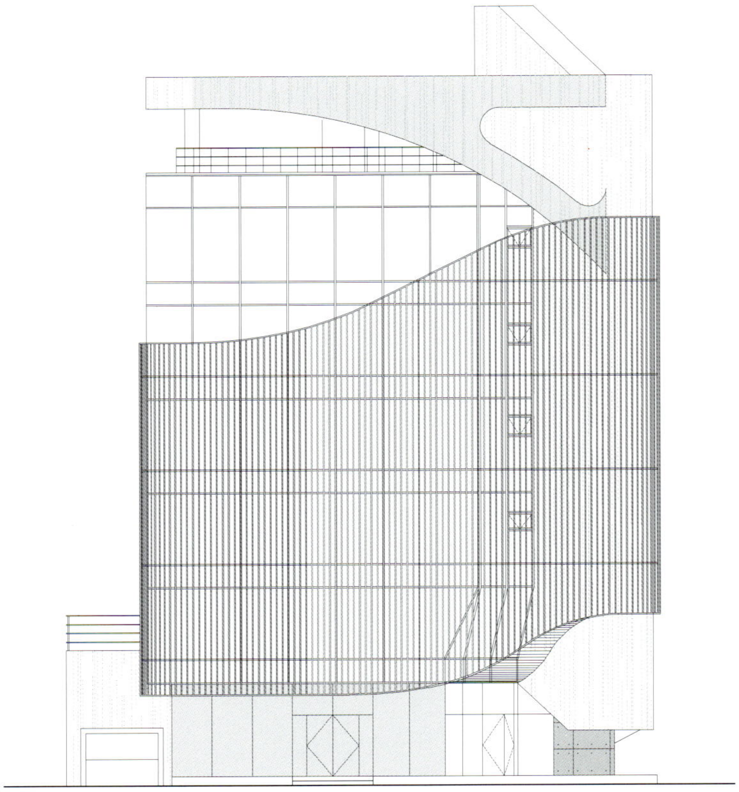

FRONT ELEVATION

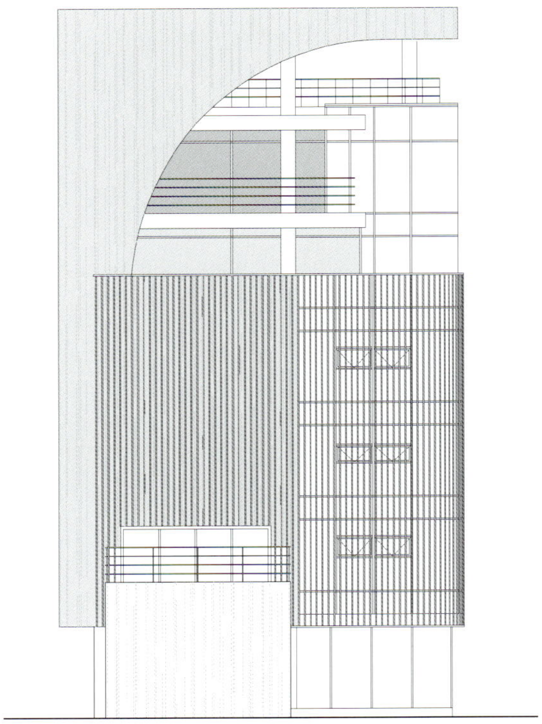

LEFT ELEVATION

VITCEN
Samsung PAVV
Samsung zipel
Samsung hauzen
Samsung SENS

SNJ OFFICE BUILDING

SNJ(주)사옥

This project is for a venture enterprise that manufactures electronic products and the client wanted a nature-friendly cultural building that accommodates gallery-like offices with creative images and multi-functions for stores. Natural ground flows including the green walking axis on the western backside and the park were connected with the building to form one big space. These spaces are neutral and natural spaces without clear boundary between 'inside' and 'outside'. Light goes into the deep inside of the building and presents open inner space. Solid walls were formed on the ground boundary in the south and the north where buildings will be located, and boards connected diagonally between both walls were set. Those board form an outdoor space and inner space surrounded those boards creates an indoor space with various topographies. Inner spaces become independent sections on each floor and the terraced outdoor garden that is connected with offices on every floor becomes a new traffic line. Various heights in stairs and light through high openings endowed variety to the space.

본 프로젝트는 전자제품을 제조하는 벤처기업사옥으로 건축주는 창의적 이미지 갤러리를 연상시키는 집무 공간과 판매점이 공존하는 다기능성, 특히 도시 속의 자연 친화 성격의 문화적 건축물을 원했다. 서측 후면의 녹지 보행 축과 공원 등 대지가 갖는 자연 흐름을 건물에 연결시켜 하나의 공간과 형태로 전환했다. 이 공간들은 더 이상 '안'과 '밖'의 경계가 분명하지 않은 중성적 자연 공간이다. 빛이 내부 깊숙한 곳까지 영역을 설정하며 개방된 내부 공간을 연출한다. 추후 건물이 들어설 남북에 접한 대지 경계면을 솔리드한 벽체로 구성하고, 양쪽의 벽체 사이에 사선으로 엇갈리며 연결되는 판들을 설정하였다. 판들은 옥외 공간으로 구성되며 판들 사이로 내부 공간이 관입되어 다양한 지형을 가진 실내 공간을 형성한다. 내부 공간은 층별로 각기 다른 독립적 영역으로 구성되고, 모든 층별 사무 영역과 이어진 계단식 외부 정원은 새로운 동선 수단이 된다. 다양한 층고, 상부 채광 등의 방식으로 공간의 다양성을 부여했다.

Architects : JNK Architects Co., Ltd(Jeon Byung Wook, Kim Myung Hyun, Park Bo Won, Kang Jin Goo, Park Chun Seok) **Location :** 1052-3 Yeongdeok-dong, Giheung-gu, Yongin, Korea **Site area :** 766.50m² **Building area :** 458.76m² **Gross floor area :** 1,996.95m² **Building coverage :** 59.85% **Floor space index :** 199.84% **Construction period :** 2009. 9 ~ 2010. 4 **Photographer :** Kim Jae Kyeong

설계 : (주)JNK종합건축사사무소(전병욱, 김명현, 박보원, 강진구, 박천석) **위치 :** 경기도 용인시 기흥구 영덕동 1052-3 **대지면적 :** 766.50㎡ **건축면적 :** 458.76㎡ **연면적 :** 1,996.95㎡ **건폐율 :** 59.85% **용적률 :** 199.84% **공사기간 :** 2009. 9 ~ 2010. 4 **시공자 :** (주)이안알앤씨 **사진 :** 김재경

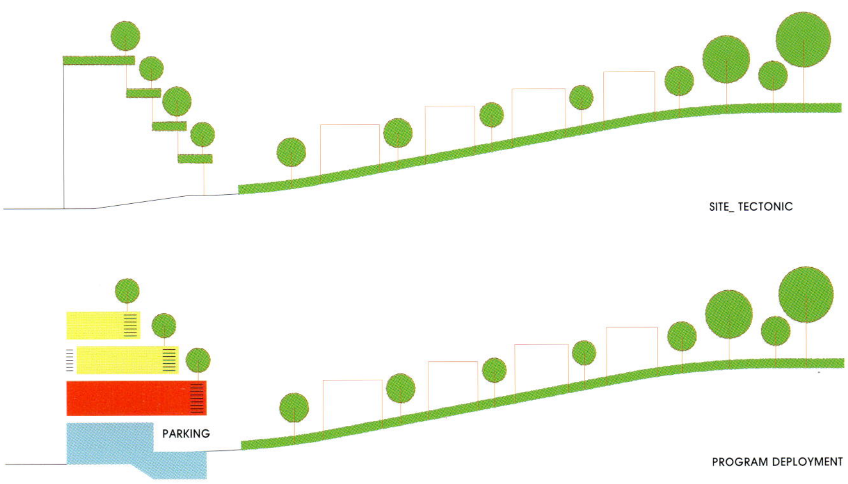

DIAGRAM

← View from East

↑ View from West ↱ Roof garden ↲ Detail of facade ↳ Extenal space of west side

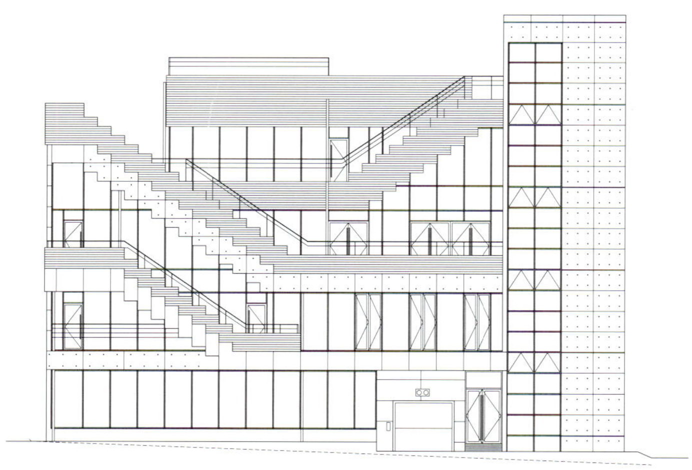

EAST ELEVATION

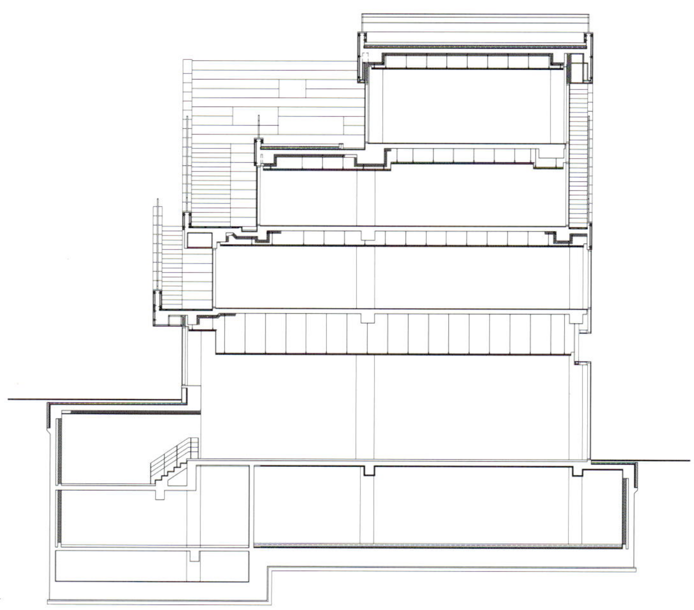

LONGITUDINAL SECTION

← Terraced outdoor garden ↑ Terraced outdoor garden

1 OPEN ABOVE
2 NEIGHBORHOOD FACILITY

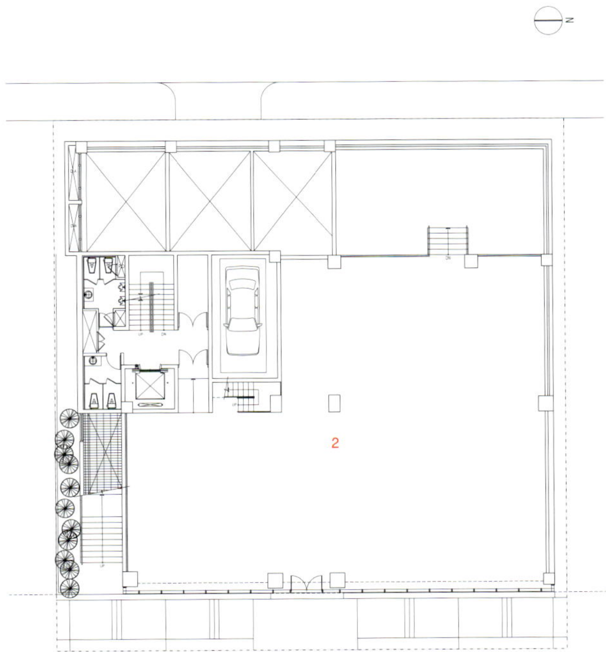

1ST FLOOR PLAN

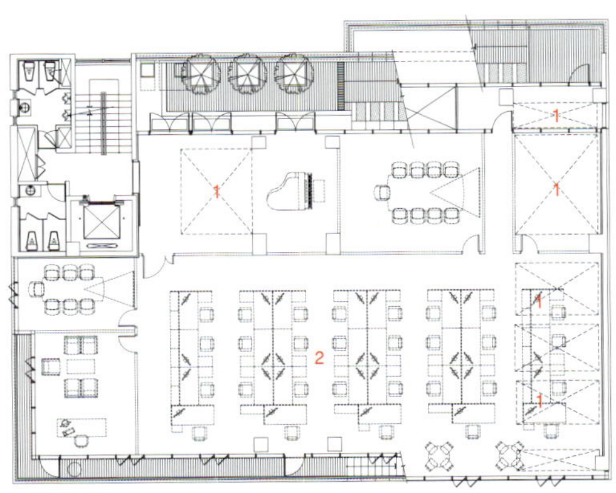

3RD FLOOR PLAN

에스엔제이주식회사
SNJ CO., LTD.

← Lobby ⌟ Meeting room ⌝ CEO room

韓醫院

DAEJEON ORIENTAL MEDICINE CLINIC & RESIDENCE

대전 한의원 주택

The client had dreamt of a building that had an oriental medical clinic on the 1st floor, home spaces for his family on the 2nd floor and a cosy, clean and special guest area where he could invite friends or relatives to have small parties or play instruments for folk music. The site was located between stores that soared unevenly around an apartment complex and an empty ground where the frivolous newness of the new town and surrealism of pointed roofs seen in fairy tales co-existed. This place is occupied with fabrications of obscure origin around a deserted ten-lane road and questionable sloped roofs and signboards made under esoteric architectural instructions on severely uneven masses adds vagueness to the area. We decided to build a rough, less polished, unfamiliar and plain house instead of a beautiful and nice looking one. We kept the external appearance original as much as possible by taking sloped roofs that come together inside and gave openness to the inner space that looked closed and stubborn through the yard on the 2nd floor.

건축주는 1층에는 직접 운영하는 한의원, 2층에는 가정집을 계획하여 가끔 친지들을 청해 술잔을 기울이며 흥이 나면 언제든 풍물판을 벌일 수 있는 작고 간결하지만 개성 있는 공간을 원했다. 대지는 아파트 주변부에 들쑥날쑥 들어선 상가와 나대지 사이, 신도시의 경박한 새로움과 동화 속 뾰족 지붕의 초현실이 공존하는 곳에 위치하고 있었다. 휑한 10차선 도로 옆으로는 출처가 불분명한 허구들이 띄엄띄엄 채워지고 요철 심한 매스 위에 근거를 알 수 없는 시행 지침이 만든 어정쩡한 경사 지붕과 간판까지 더해졌다. 우리는 아름답거나 멋있는 집이 아니라, 오히려 투박하고 덜 세련된, 낯설고 못생긴 집을 짓기로 했다. 안쪽으로 모이는 경사지붕을 취하여 밖으로는 최대한 원초적인 형태를 유지하고, 고집스레 막힌 듯 보이지만 2층 마당을 통해 개방적인 내부를 계획했다.

Architect : Dia, Architecture Co.(Chung Hyun A) **Location :** 657-1 Banseok-dong, Yuseong-gu, Daejeon, Korea **Site area :** 246.7m² **Building area :** 134.5m² **Gross floor area :** 247.6m² **Construction period :** 2008. 9 ~ 2009. 2 **Photographer :** Park Wan Soon, Park Chan Woo

설계 : 디아건축(정현아) **위치 :** 대전광역시 유성구 반석동 657-1 **대지면적 :** 246.7㎡ **건축면적 :** 134.5㎡ **연면적 :** 247.6㎡ **공사기간 :** 2008. 9 ~ 2009. 2 **사진 :** 박완순, 박찬우

← Overall view

崔光洵
韓醫院

← View from West ↑ View from North ↓ View from South

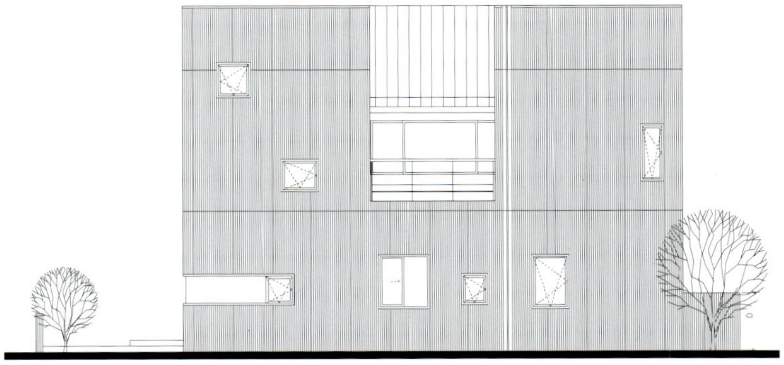

SOUTH ELEVATION

WEST ELEVATION

← Deck on the 2nd floor
↱ Deck on the 2nd floor
↳ View of the dining room from the deck

1 LIVING ROOM
2 DECK
3 BEDROOM
4 DOCTOR'S OFFICE
5 CARE SPACE
6 RECEPTION

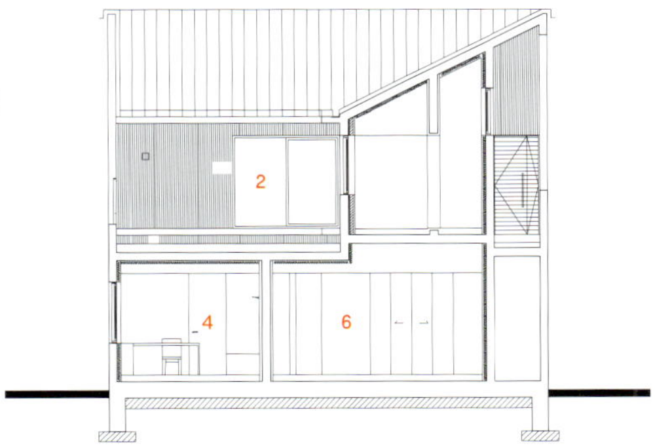

SECTION

SECTION

↑ View of the dining room ↓ View of the dining room with mezzanine floor

↰ Stairs ↲ Stairs ↑ View of the mezzanine

1 RECEPTION
2 DOCTOR'S OFFICE
3 CARE SPACE
4 DECK
5 DINING ROOM
6 LIVING ROOM
7 BEDROOM

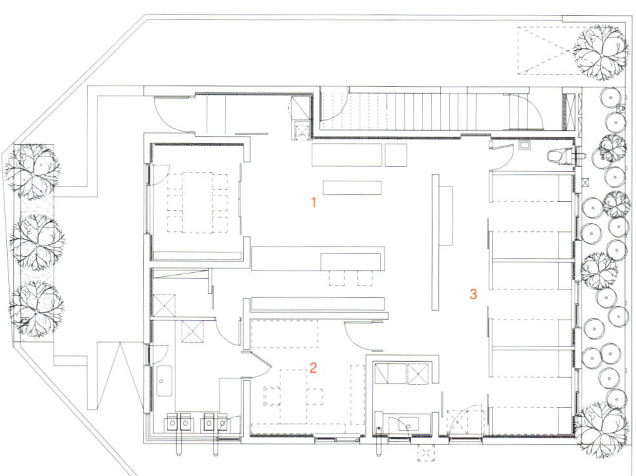

1ST FLOOR PLAN

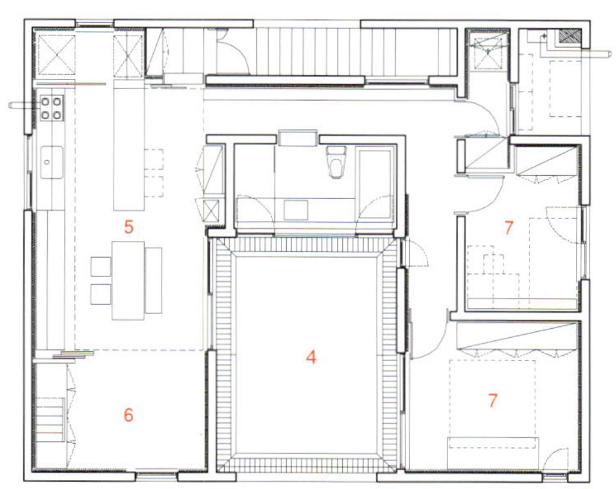

2ND FLOOR PLAN

⌐ View of the reception area ⌟ View from the reception area to a tearoom ↑ View of the reception area

P

DEMIAN BUILDING

데미안 빌딩

It is very easy to see stores fully covered with signboards in Seoul. The site of Demian building is located in Junggye-dong, one of typical apartment complexes in Seoul, and its nearby stores also display a messy urban landscape with a facade created by huge numerous signboards. Among familiar but distressing landscapes they wanted to build a house in a new fresh way. Commercial desire of store lessors as well as desire to create advisable relation between a city and buildings were reflected in the design of Demian building. Through such investigation, Demian building became materialized in a way that the house is recognized as one volume rather than a frame of horizontal lines and lattices. The group of irregular concrete vertical lines that play two roles of a structure and a decorative skin break down typical structure forms and allow the house to be recognized as one volume. The canopy that was added to the upper part of the 1st floor like a cornice line gives proper scale to the pedestrian level and defines a section for a signboard clearly. It also expands the building border and induces the urban flow into the building. The main entrance in a height of two stories responds to nearby urban scale as well as shows welcoming gesture to visitors.

Architect : Kim Seung Hoy(Seoul National University) + KYWC Architects(Kang Won Phil) **Location :** Junggye-dong, Nowon-gu, Seoul, Korea **Site area :** 647.00m² **Building area :** 382.42m² **Gross floor area :** 1,367.98m² **Building coverage :** 59.1% **Floor space index :** 194. 08% **Structure :** RC, SC **Exterior finishing :** Exposed concrete, Curtain wall **Construction period :** 2008. 6 ~ 2009 .3 **Photographer :** Kim Jae Kyeong

설계 : 김승회(서울대학교) + 경영위치건축사사무소(강원필) **위치** : 서울특별시 노원구 중계동 157-5번지 외 9필지 **대지면적** : 647.00㎡ **건축면적** : 382.42㎡ **연면적** : 1,367.98㎡ **건폐율** : 59.1% **용적률** : 194.08% **구조** : 철근콘크리트조, 철골조 **외부마감** : 노출콘크리트, 커튼월 **공사기간** : 2008. 6 ~ 2009.3 **사진** : 김재경

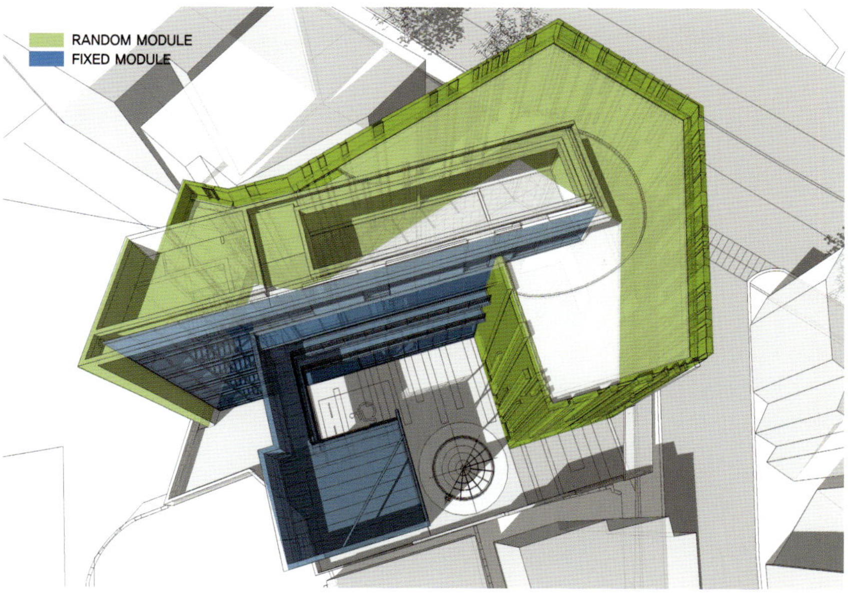

RENDERING IMAGE

← View from East

← View from the street

WINDOW MODULE

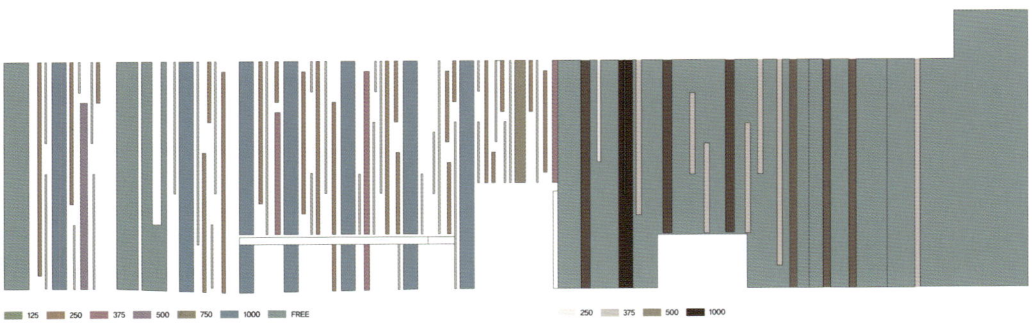

ELEVATION MODULE

간판으로 가득 채워진 상가들의 모습은 서울에서 매우 낯익은 풍경이다. 아파트 단지와 상가의 간판이 구성하는 파사드로 서울의 전형적인 풍경을 보여주는 중계동에 데미안 빌딩이 위치한다. 낯익지만 괴로운 풍경들 사이에 새로운 방식으로 집을 창조하고 싶었다. 상가 임대인들의 상업적 욕구를 반영하면서 동시에 바람직한 도시와 건축의 관계를 만드는 방식을 설계 과정에 반영하였다. 이러한 탐구를 통해서 건물이 인식되는 형태를 수평선과 격자의 프레임이 아닌, 하나의 볼륨으로 구체화하였다. 구조와 장식적 스킨, 이 두 가지 역할을 동시에 감당하는 불규칙한 콘크리트 수직선의 집합은 상식적인 구조 형식을 무너뜨리면서 하나의 볼륨으로 건물을 인식하게 한다. 1층 상부에 코니스 라인처럼 덧붙여진 캐노피는 보행자 레벨에 적절한 스케일을 부여하는 동시에 간판을 붙일 영역을 분명히 정의한다. 그리고 건물의 경계를 확장시키고 도시의 흐름을 내부로 끌어오는 역할도 수행한다. 두 층 높이의 주출입구는 주변의 도시 스케일과 반응하면서 방문자들을 환영하는 제스처를 보여준다.

The two level height main entrance / View from the bridge

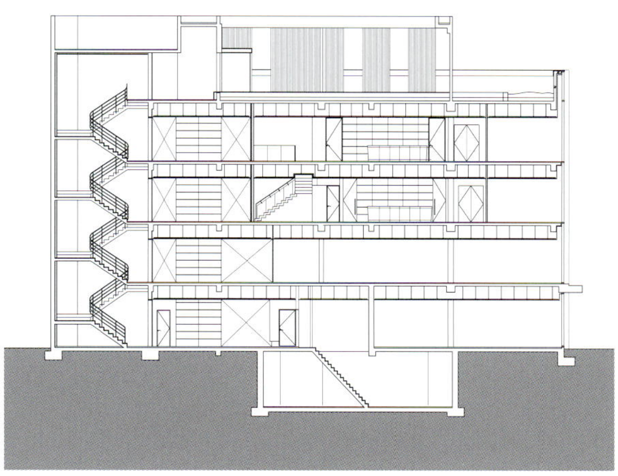

SECTION

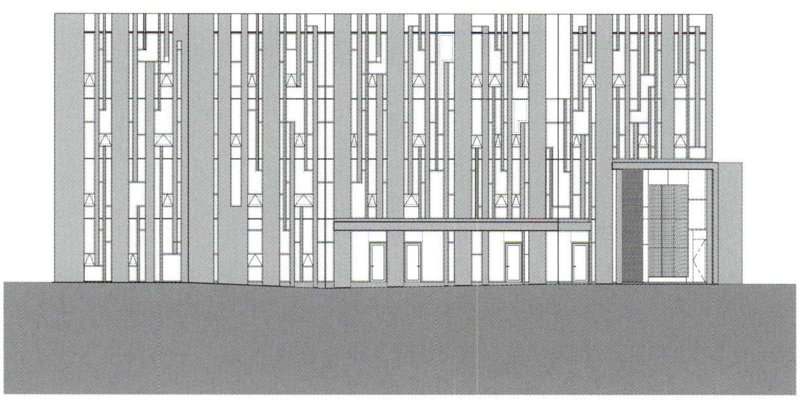

ELEVATION

1 RENTAL SPACE
2 MAIN ENTRANCE
3 MEDICAL EXAMINATION CENTRE
4 DOCTOR'S OFFICE
5 LABORATORY

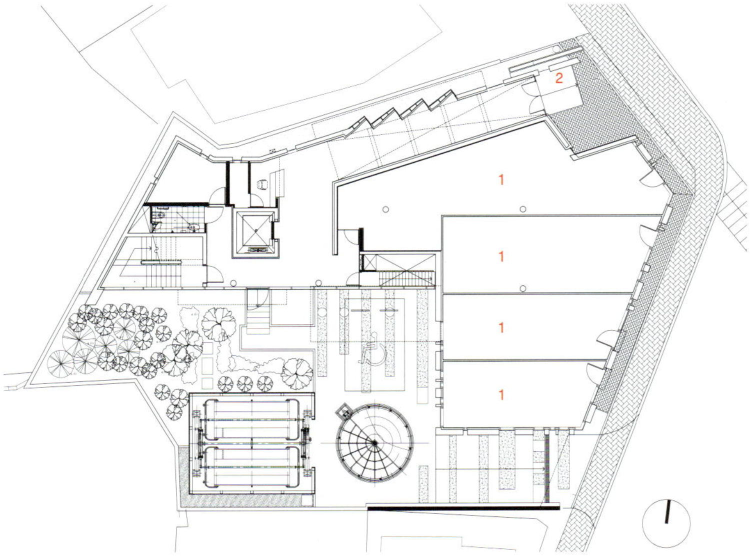

1ST FLOOR PLAN

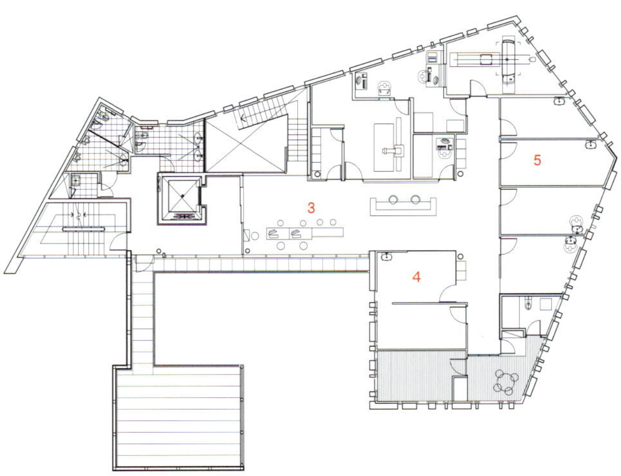

4TH FLOOR PLAN

View of parking tower from terrace / View from the rooftop of parking tower / Interior view of the 3rd floor / Corridor of the 1st floor

MONO TOWER

모노타워

Mono Tower, an IT company building is located behind the boulevard near Sunreung. The construction was restrained by the diagonal limit of the 6m-long roads on both sides of the site which is a corner-shaped slope while the right-to-have-sunlight of the site was also restricted as it faces the North. However, our work was not bothered by the restrictions since we originally planned to build a building that makes a harmony with the surroundings. We also considered the spatial programs as well as the floor plan as the building should accommodate many people. To emphasize the originality of the IT company building, we finished the building with exposed concrete and light curtain walls. In addition, circles and oblique lines were introduced to the volume in order to induce dynamic and soft images unlikely the surrounding box-shaped buildings. In consideration of the demand of lease, we organized the spaces. Also, we planned a roof garden to create a refreshing working environment and secured sufficient sunshine in the basement through a sunken garden.

모노타워는 IT 관련 사옥으로 선릉 근처의 대로변 바로 뒤편 위치한다. 코너 형태의 경사지인 대지 양쪽으로 6m 도로가 인접하여 도로 사선 제한의 영향을 받으며, 북쪽으로 면한 대지로 일조권의 제한도 받았다. 이러한 제약과는 상관없이 원래부터 이러한 자태로 주변 컨텍스트와 조화되는 건물로 만들고자 했다. 건물 자체에 수용해야 할 인원이 많아서 평면 계획은 물론 공간 프로그램도 고려했다. IT 사옥의 독창성을 살리기 위해 주물성이 좋은 노출콘크리트와 맑고 밝은 느낌의 커튼월 시스템으로 마감했다. 또한 주변의 박스형의 건물들과는 다르게 역동적이면서 부드러운 이미지를 유도하기 위해 원호와 사선을 적극적으로 사용하여 볼륨 계획에 도입했다. 근린생활시설인 만큼 임대수요를 고려하여 공간을 구성했다. 또한 쾌적한 업무환경 조성을 위해 옥상정원을 계획했으며, 선큰 가든으로 지하 공간에 충분한 일조량을 확보했다.

Architect : ARCHIDREAM ARCHITECTS & ENGINEERS (Song Choon Sik, Yang Seung Yeol, Kim Kil Soo, Yoon Jae Min, Park Myung Jin) **Location :** Yeoksam-dong, Gangnam-gu, Seoul, Korea **Site area :** 258.80m² **Building area :** 154.36m² **Gross floor area :** 720.91m² **Building coverage :** 59.64% **Construction period :** 2009. 6 ~ 2010. 5 **Photographer :** Lim Ji Won, Kim Min Soo

설계 : (주)아키드림 건축사사무소(송춘식, 양승열, 김길수, 윤재민, 박명진) **위치 :** 서울특별시 강남구 역삼동 695-30 **대지면적 :** 258.80㎡ **건축면적 :** 154.36㎡ **연면적 :** 720.91㎡ **건폐율 :** 59.64% **공사기간 :** 2009. 6 ~ 2010. 5 **사진 :** 임지원, 김민수

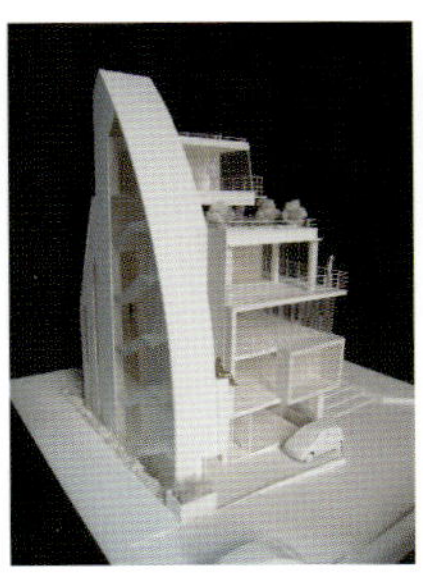

MODEL IMAGE

← The complete view

RAMADA SEOUL

← Overall view
→ View of the front right side
↳ View of the right side

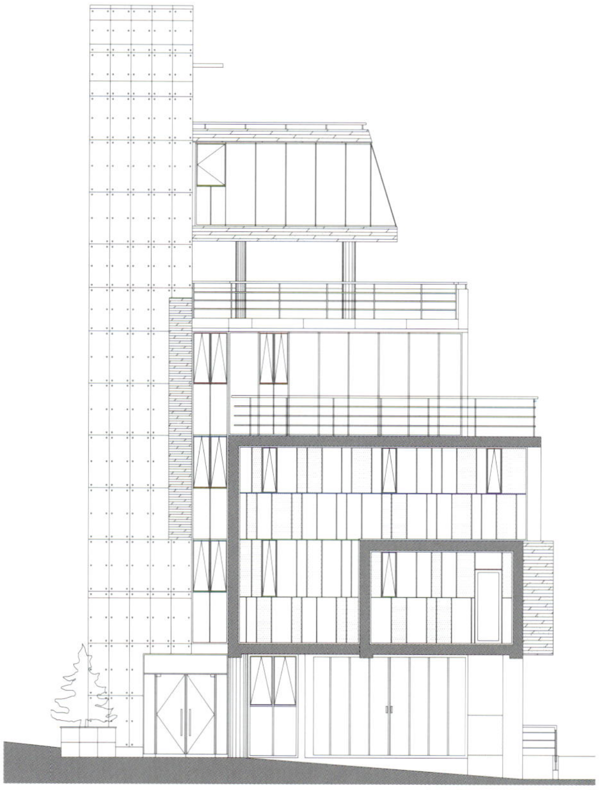

FRONT ELEVATION

RIGHT ELEVATION

← View of the back right side ↑ Interior view from below

1 OFFICE
2 NEIGHBORHOOD FACILITY
3 SUNKEN GARDEN
4 ROOF GARDEN

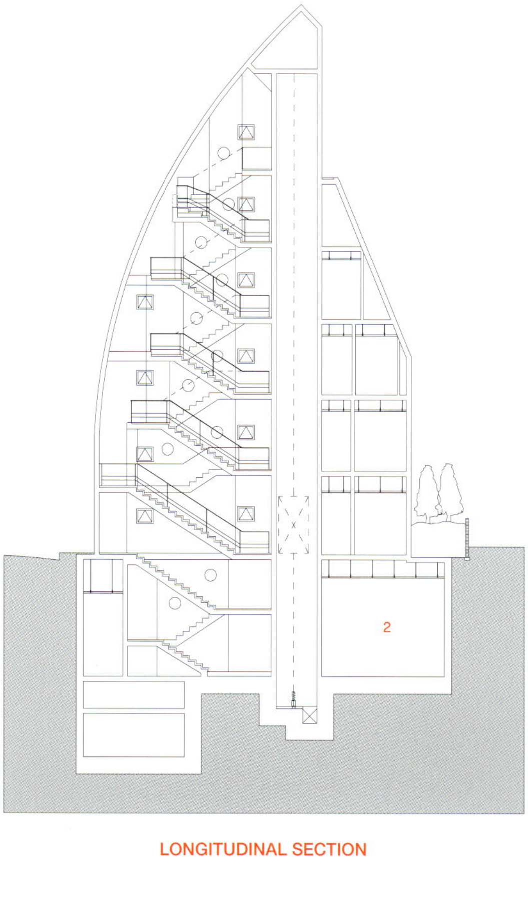

LONGITUDINAL SECTION

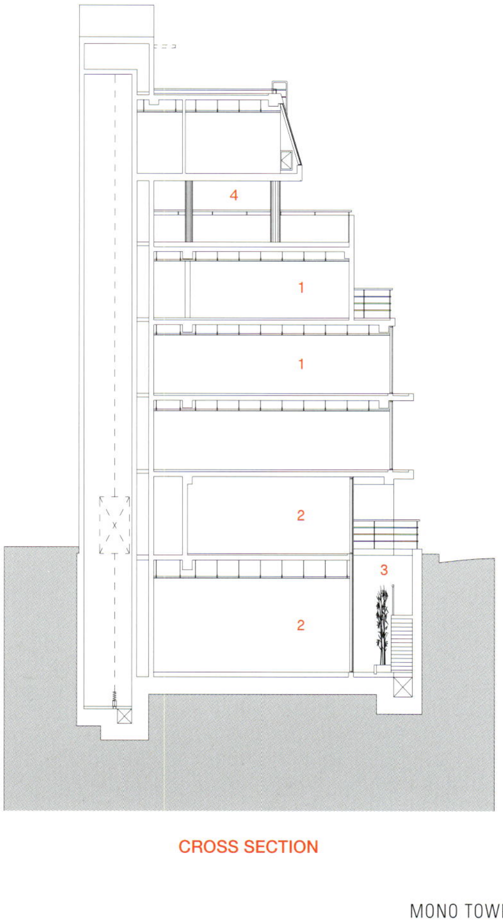

CROSS SECTION

← Night exterior view
┌ Stairs
└ Stairs

↑ Roof garden ↱ Roof garden ↳ Roof garden

1 OFFICE
2 HALL
3 NEIGHBORHOOD FACILITY
4 OPEN
5 PARKING
6 ENTRANCE

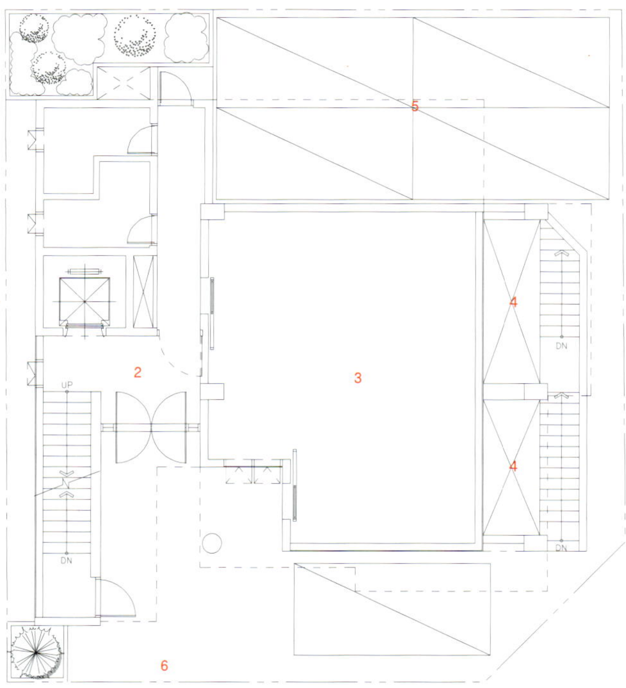

1ST FLOOR PLAN

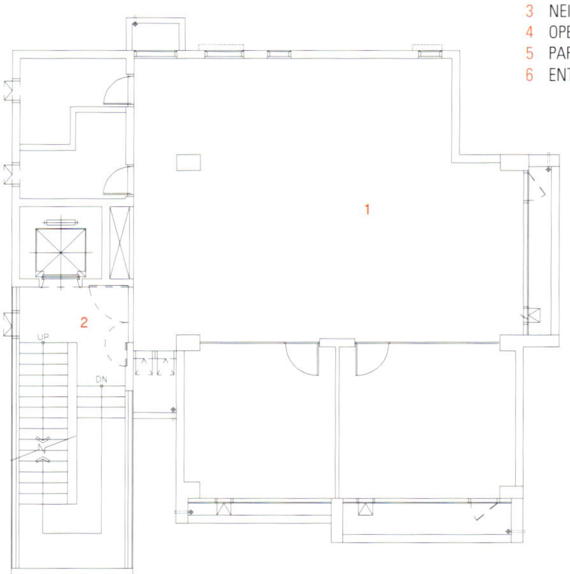

2ND FLOOR PLAN

gelato & waffle
caffé bene
카페베네
AURORA
426
AURORA

AURORA BUILDING

오로라 빌딩

The client requested a building in a unique shape on the roadside. We designed considering of a unique space apart from dry office buildings more than a unique shape. We intended to building an environment-friendly building in consideration of southward sunshine, natural ventilation and green for a rest(such as a garden). To induce enough sunshine, optimize ventilation and make a space to contain nature, the core was spread out and rearranged. Sunken and a wall fountain were introduced in consideration of underground conditions. The main concept was movement and the purpose was to make an alive space. Variable spaces were intensely arranged along somewhat long traffic lines to create a lively space. The courtyard is surrounded by the divided core and the rhythmically arranged bridge and variable nature is induced into the building through the courtyard to enrich movement within the building. To add movement to the mass, it was twisted. Since the theme was movement, naturally dynamic shapes were formed. As a psychological environment and a device of movement, bamboo was induced for the front of the mass. To make piled masses look like individually moving parts, it was necessary to settle details and structures of twisted masses, which was the most difficult process.

건축주는 대로변에서 눈에 띄는 독특한 모양을 갖춘 건물을 요구했다. 기존의 오피스 건물의 건조한 공간을 탈피한 독특한 공간을 구성하고자 했다. 남향의 일조, 자연 통풍, 휴게 녹지(정원) 등의 조건을 살릴 친환경적인 건물을 계획했다. 일조와 환기를 고려하여 코어를 분리하고, 이격시켜 충분한 채광이 유입되는 자연을 담는 공간을 만들었다. 지하층의 환경을 고려하여 선큰과 벽천을 도입했다. 주된 개념은 움직임(動)이며 목적은 살아있는 공간을 만드는 것이었다. 다소 긴 동선과 그 동선에 따라 변화하는 공간들을 긴장감 있게 배치하여 생동감을 부여했다. 코어의 분리와 리드미컬한 브리지의 배치로 위요된 중정을 만들고, 중정을 통해 자연의 변화를 건물 내부로 끌어들임으로써 움직임은 더욱 풍성해진다. 매스를 비틀어 역동적인 움직임을 도입하였다. 움직임을 기본 개념으로 진행하여 실제 움직이는 듯한 형태는 자연스럽게 도출되었다. 심리적 환경성과 움직임의 장치로 매스 전면부에 대나무를 도입했다. 쌓여진 매스가 각각 따로 움직인 듯한 형태를 위하여 틀어진 매스가 겹쳐지지 않도록 디테일을 풀고 구조를 해결하는 것이 어려움이었다.

Architect : BUMJIN ARCHITECTS & ENGINEERS(Cho Kwang Yeon, Kim Chang Seok, Yoo Dae Jong) **Location** : 120-12 Samseong-dong, Gangnam-gu, Seoul, Korea **Site area :** 551.30m² **Building area :** 270.57m² **Gross floor area :** 2,303.36m² **Building coverage :** 49.35% **Floor space index :** 247.95% **Photographer :** Kim Jae Kyeong, BUMJIN ARCHITECTS & ENGINEERS

설계 : (주)범진종합건축사사무소(조광연, 김창석, 유대종) **위치** : 서울특별시 강남구 삼성동 120-12 **대지면적** : 551.30㎡ **건축면적** : 270.57㎡ **연면적** : 2,303.36㎡ **건폐율** : 49.35% **용적률** : 247.95% **사진** : 김재경, (주)범진종합건축사사무소 제공

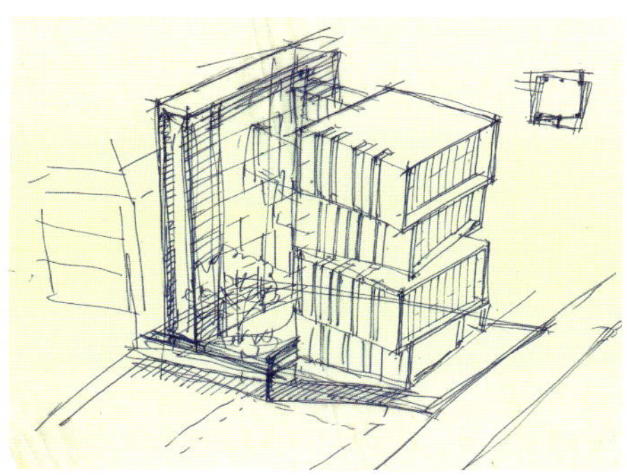

SKETCH

← General view

Exterior view · Night exterior view · View of main facade · Night Exterior view

↑ Overall view showing the urban context

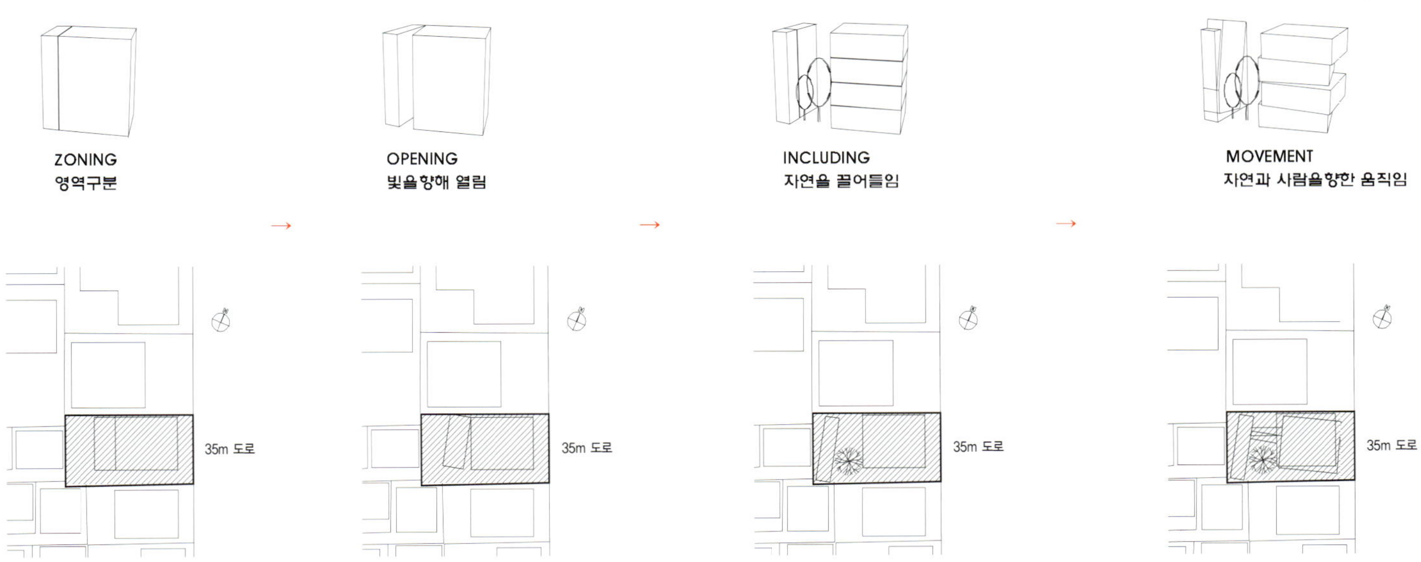

PROCESS DIAGRAM

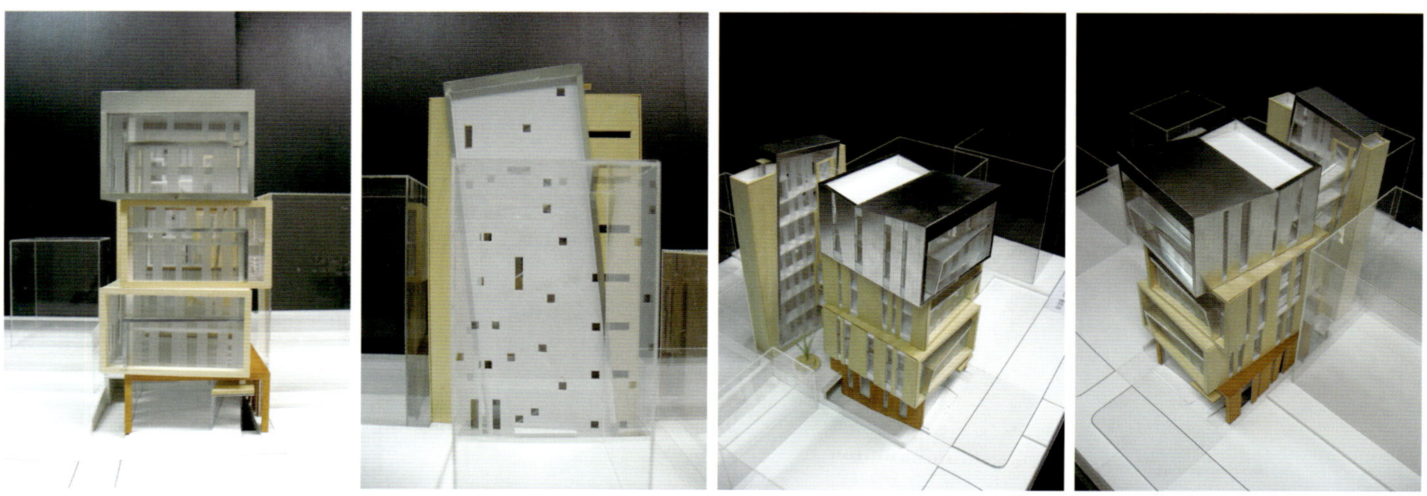

MODEL STUDY

← View from the left side ↑ Series of bridges viewed from courtyard

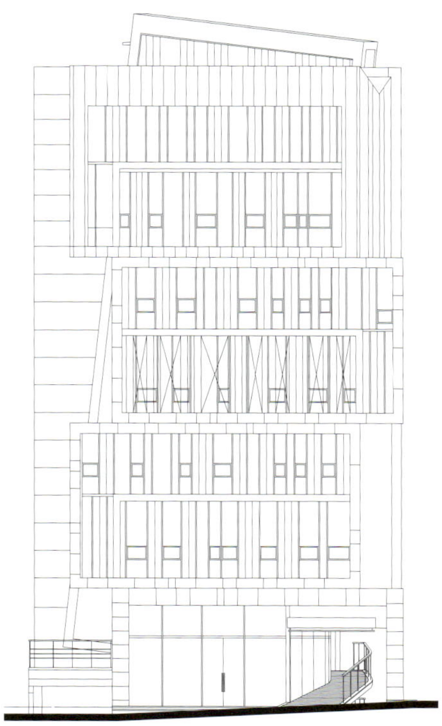

FRONT ELEVATION

LEFT ELEVATION

caffe bene

← View from the main entrance to B1 floor ↰ Part of the facade ↲ Window ↑ Courtyard

1 NEIGHBORHOOD FACILITY
2 STAIRS BUILDING
3 OFFICE AREA
4 BRIDGE

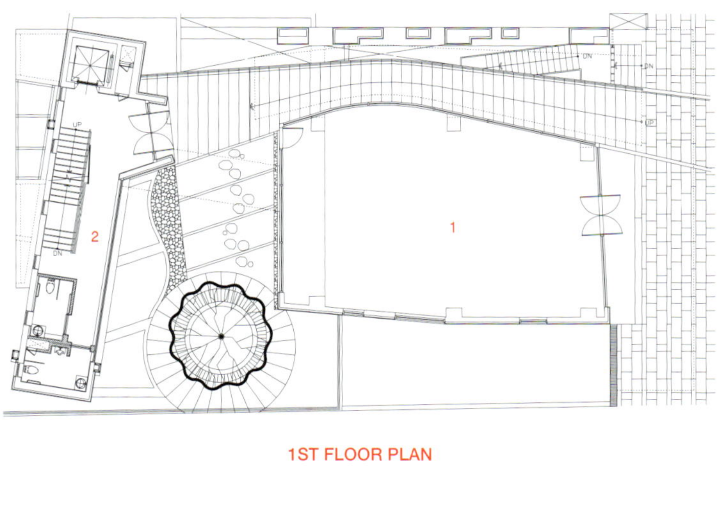

1ST FLOOR PLAN

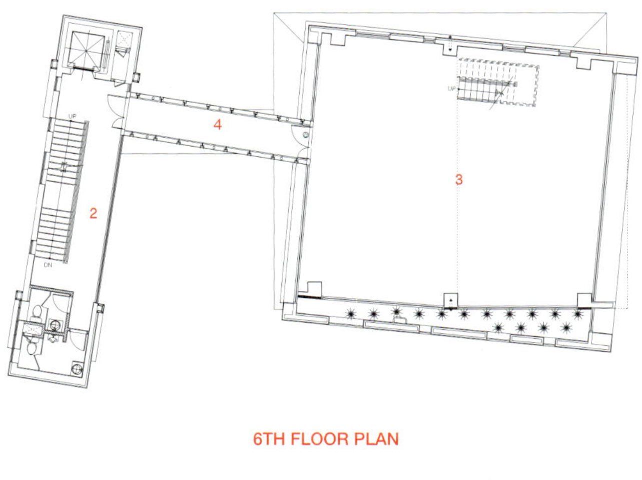

6TH FLOOR PLAN

orang
cafe
bistro
187
cafe & bistro
orangrang

NORANG GALLERY

노랑갤러리

The narrow and long site is in the middle part of street in the urban center. Norang Gallery was planned as a small culture complex located along the street on where workers and residents of the area enjoy relaxing time. To take advantage of the river and the greens, a clear and transparent(glass) facade was suggested to enjoy the nature from everywhere in the buildings. Only the front was open while the other three sides were closed by the monotone concrete walls in order to maximize visibility. By inserting a skylight into the mass which was irregularly layered in accordance with the legal regulations applied on the site which is adjacent to the main road in three sides, different volumes and depth between the floors were created in the uniform space. The ground floor was made as an open buffering space which can stimulate eyecontact with the pedestrians. The cafe gallery on the 1st gallery has the ample height created by the crossed skip-floors of two units. For the right-to-have sunlight, the volume was vertically divided into six masses. The mass, cut along the two sides of the site, was connected to three plots again, depending on its spatial structure and function. The various flows of the circulation between the floors that were made of different-size rectangle layers is the main concept of Norang Gallery. In the relation with the metasequoia street, the stairway projected in the building is irregularly connected and rises to the top. Norang Gallery meets another new potential space via this tapering stairway in zigzag. It suggests a function as an horizontal gallery not only as a vertical passage. It is expected that the space will function as a spiral gallery which provides another possibility out of the typical stairways.

Architect : STUDIO KOOSSINO(Koo Seung Min) **Location :** 454-14 Dogok-dong, Gangnam-gu, Seoul, Korea **Site area :** 264.60m² **Building area :** 157.85m² **Gross floor area :** 725.96m² **Building coverage :** 59.66% **Floor space index :** 198.17% **Design period :** 2007. 3 ~ 2007. 8 **Photos offer :** STUDIO KOOSSINO

설계 : 건축스튜디오 꾸씨노(구승민) **위치** : 서울특별시 강남구 도곡동 454 - 14 **대지면적** : 264.60㎡ **건축면적** : 157.85㎡ **연면적** : 725.96㎡ **건폐율** : 59.66% **용적률** : 198.17% **설계기간** : 2007. 3 ~ 2007. 8 **사진** : 건축스튜디오 꾸씨노 제공

대지는 도심의 가로수 길의 중앙부에 폭이 좁고 깊이가 긴 직사각형의 세장비로 위치한다. 인근 직장인이나 동네 주민들이 가볍게 산책을 즐기는 길을 따라 작은 복합문화공간의 쉼터로 계획하였다. 양재천과 도심의 가로 숲이라는 이점을 최대한 살려 내부의 어느 곳에서든 자연을 만끽할 수 있도록 전면을 투명한 스킨(유리벽)으로 제안하였다. 시계(視界)를 극대화하기 위해 무채색 콘크리트로 3면의 벽을 닫아 오로지 전면만을 소통하게 하였다. 일면의 도로와 3면에 접한 대지의 법적 용적에 따라 서로 다르게 쌓아 올린 매스에 천창을 삽입하여 획일화 된 공간을 층간의 변화를 통해 다양한 볼륨과 깊이를 유도하였다. 지상층은 지나는 사람들과 시선의 소통을 꾀할 수 있는 완충 공간으로서 개방성을 고려하였다. 1층 카페 갤러리는 두 유닛의 스킵 플로어로 맞물려 충분한 층고를 확보하였다. 일조권을 고려하여 볼륨을 수직으로 6개로 분절하였다. 대지의 두면을 따라 지상으로 삭감된 매스의 흐름은 다양한 공간구조와 기능에 따라 다시 세 개의 플롯으로 연계하였다. 각기 다른 직사각형이 쌓아올려진 플로어와 플로어 사이 다양한 동선의 흐름이 주된 개념이다. 메타세콰이아 가로수 길과 연계하여 건물 속에 투영된 계단은 불규칙적으로 이어져 끊임없이 상층부로 향한다. 갤러리는 지그재그로 좁혀져 가는 계단을 따라 또 하나의 새로운 잠재적 공간을 만나게 하였다. 단지 수직 동선의 이동통로가 아닌 가로형 갤러리로서의 기능을 제안하였다.

SKETCH

← View of the front facade

← Entrance ↱ View of the front left side → View of the front right side

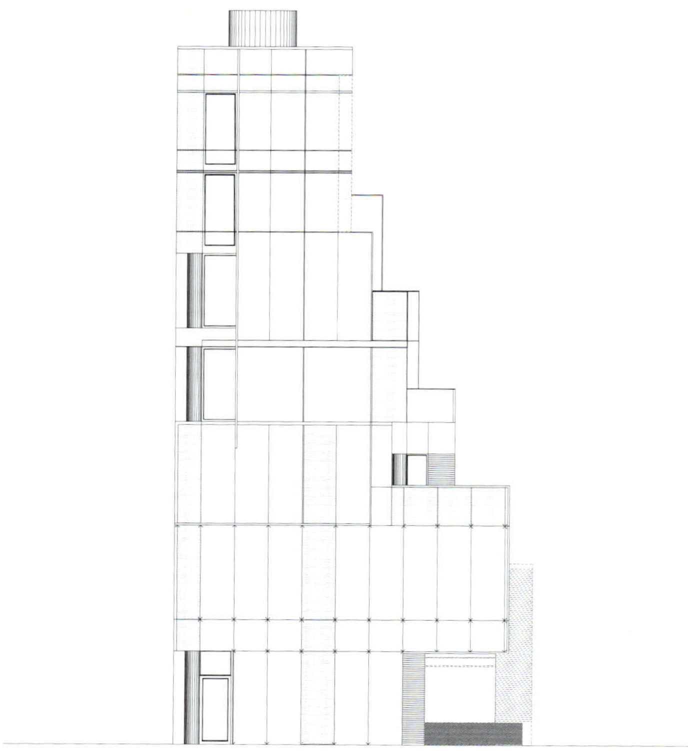

FRONT ELEVATION

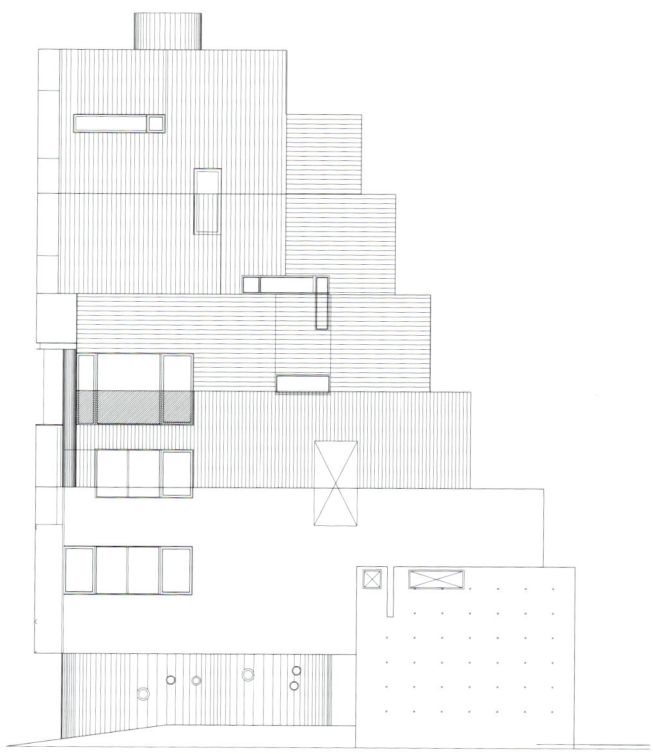

RIGHT ELEVATION

gallery norang
cafe
bistro
gallery cafe & bistro
norang
story'n kids

Interior view of the 6th floor

↑ Stairs

↑ Cafe ↓ Restaurant ↱ Circular stairs ↳ Interior view

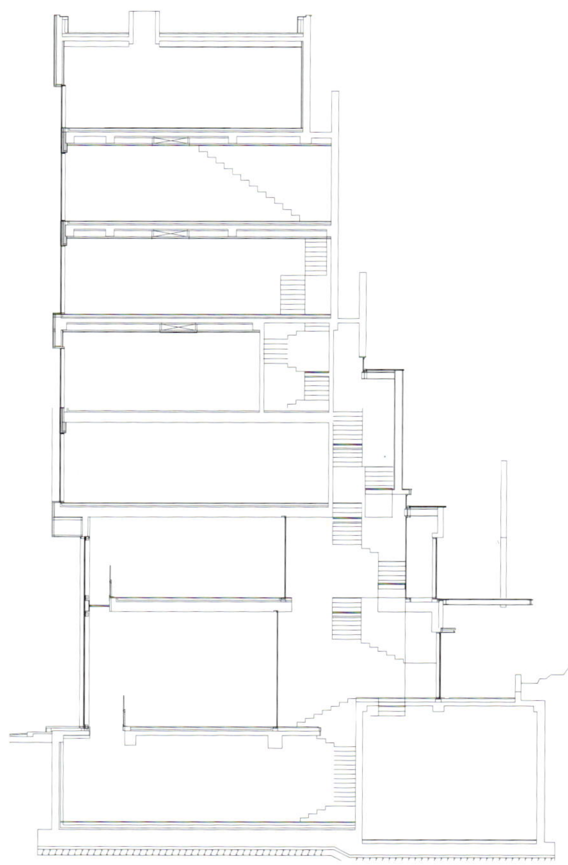

SECTION

Cafe Restaurant Gallery

1 CAFE GALLERY
2 RESTAURANT

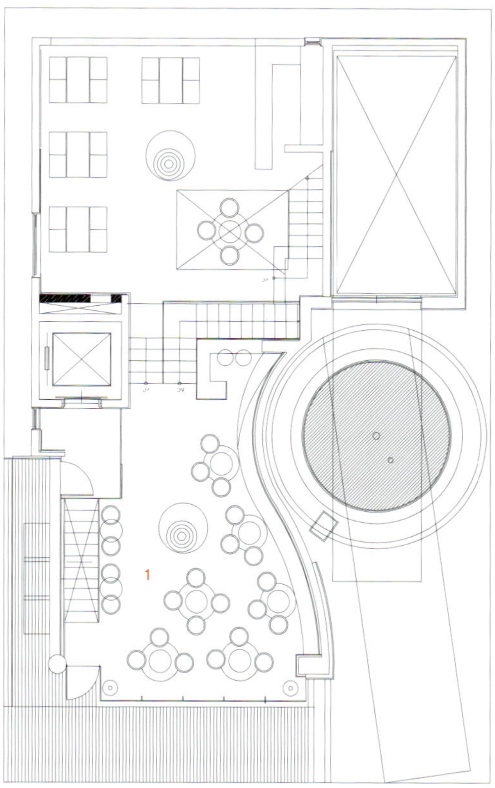

1ST FLOOR PLAN

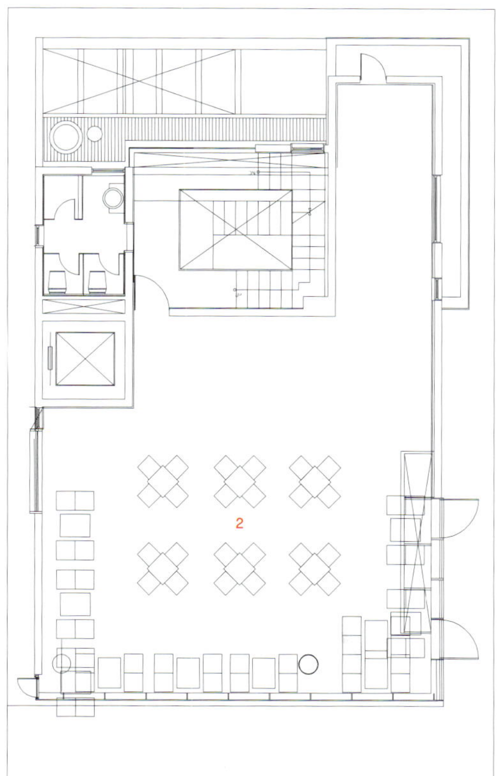

2ND FLOOR PLAN

FTOWN BUILDING

F타운 빌딩

In designing tenant buildings for which the interiors will be determined afterwards, rather than defining the spaces it is important that there is a wide spatial potential. For example, the freedom to build walls in any arrangement is more important than how the walls are actually arranged. A typical tenant building has a tree-like organization in which each floor is independently accessed by elevator from the entrance at the first floor. With regard to this project, in addition to the normal circulation, we proposed using voids to connect the spaces of each floor in a spiral arrangement, incorporating a loop in the unidirectionally ordered structure. The access routes to each space are thereby multiplied other possibilities for grouping the spaces arise, enabling integrated usages or tenants that cover several floors. Additionally, the four variations in floor height and the flexibly organized mechanical services allow a variety of spatial choices.

내부 공간의 형태가 차후에 결정되는 임대 건물 계획에서 여유 공간의 잠재성 문제는 공간을 정의하는 것보다 더욱 중요하다. 가령, 벽의 가변성과 유동성은 실제 벽체의 배치 보다 중요하다. 전형적인 임대 건물은 나뭇가지가 뻗어 나가듯이 1층 입구에서부터 엘리베이터를 통해 각 층에 독립적으로 접근이 가능하다. F타운 빌딩은 이러한 일반적인 동선 체계와 함께 보이드 공간을 활용하여 하나의 고리와 같은 구조로 각 층의 공간을 나선형의 배치로 연결시켰다. 각 공간으로 접근하는 방법이 다양화 되고, 여러 층의 통합된 활용 혹은 거주를 가능하게 하는 공간 그룹화의 다른 방법들도 가능하게 되었다. 뿐만 아니라, 네 가지 유형의 다양한 층 높이와 유연하게 조직된 설비 시설들은 공간 선택의 다양성을 제공한다.

Architect : Atelier Hitoshi Abe(Hitoshi Abe) **Location :** Tsutsujigaoka, Miyagino-ku, Sendai, Japan **Site area :** 376.96m² **Building area :** 324.59m² **Gross floor area :** 1,976.88m² **Construction period :** 2006. 8 ~ 2007. 7 **Photographer :** Daici Ano

설계 : 아틀리에 히토시 아베(히토시 아베) **위치 :** 일본 센다이 미야기노구 츠츠지가오카 **대지면적 :** 376.96㎡ **건축면적 :** 324.59㎡ **연면적 :** 1,976.88㎡ **공사기간 :** 2006. 8 ~ 2007. 7 **사진 :** 다이시 아노

PROCESS DIAGRAM

← Night view from Southwest

1 TENANT AREA
2 TERRACE
3 VOID SPACE

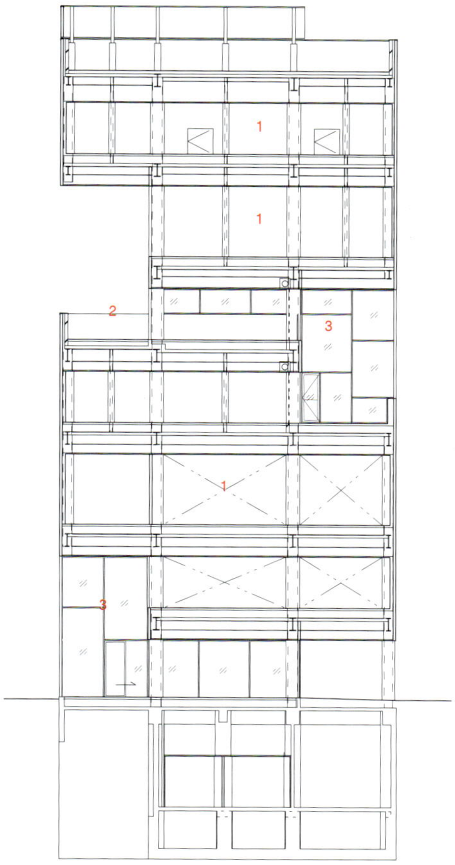

SECTION

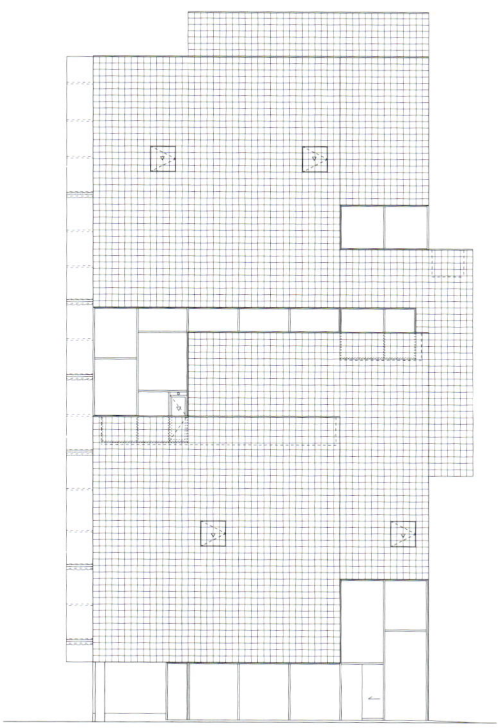

WEST ELEVATION

← View from Southeast ┌ Exterior view of protruding volume
└ Protruding volume as seen from below

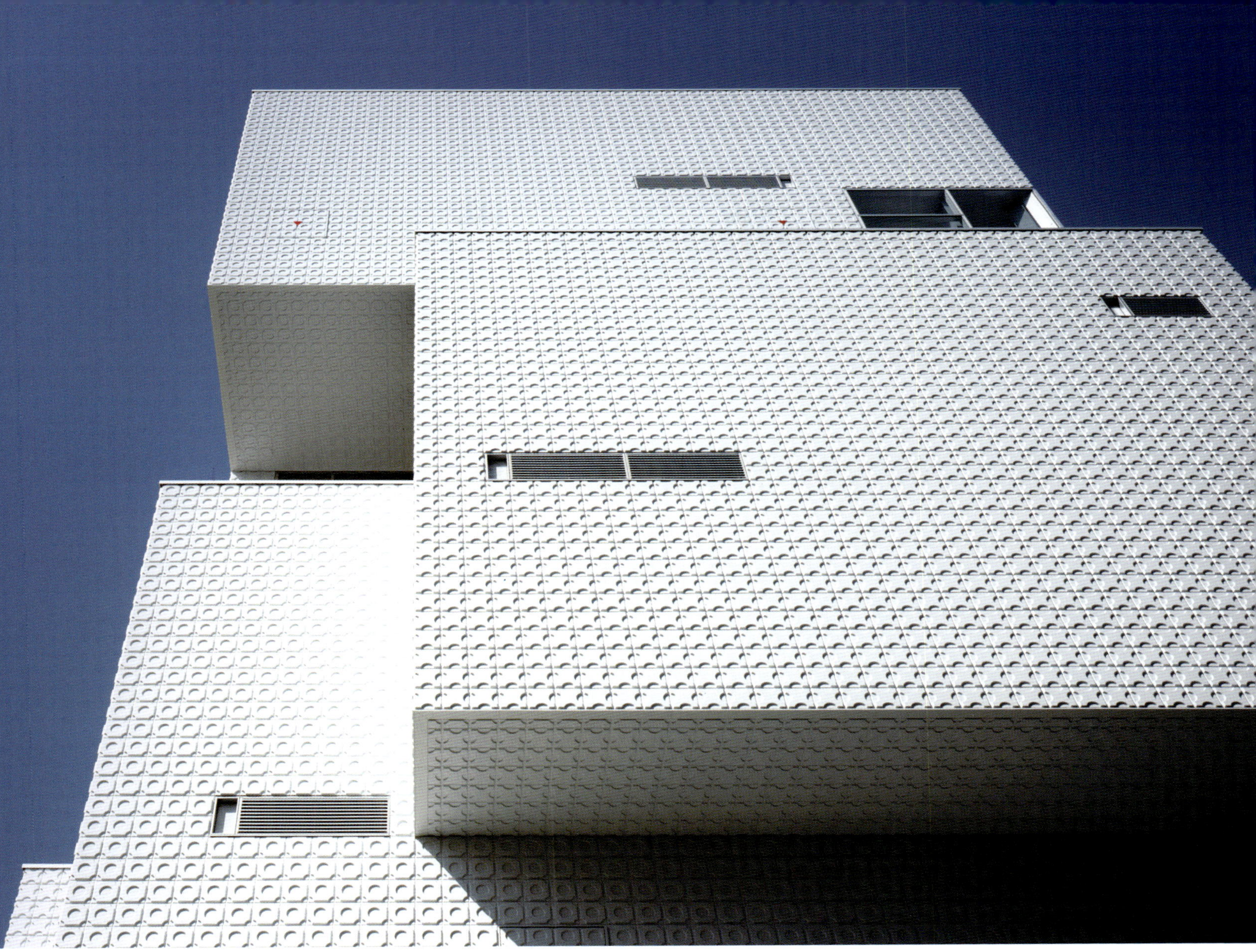

← View from West ↑ Detail of the South facade

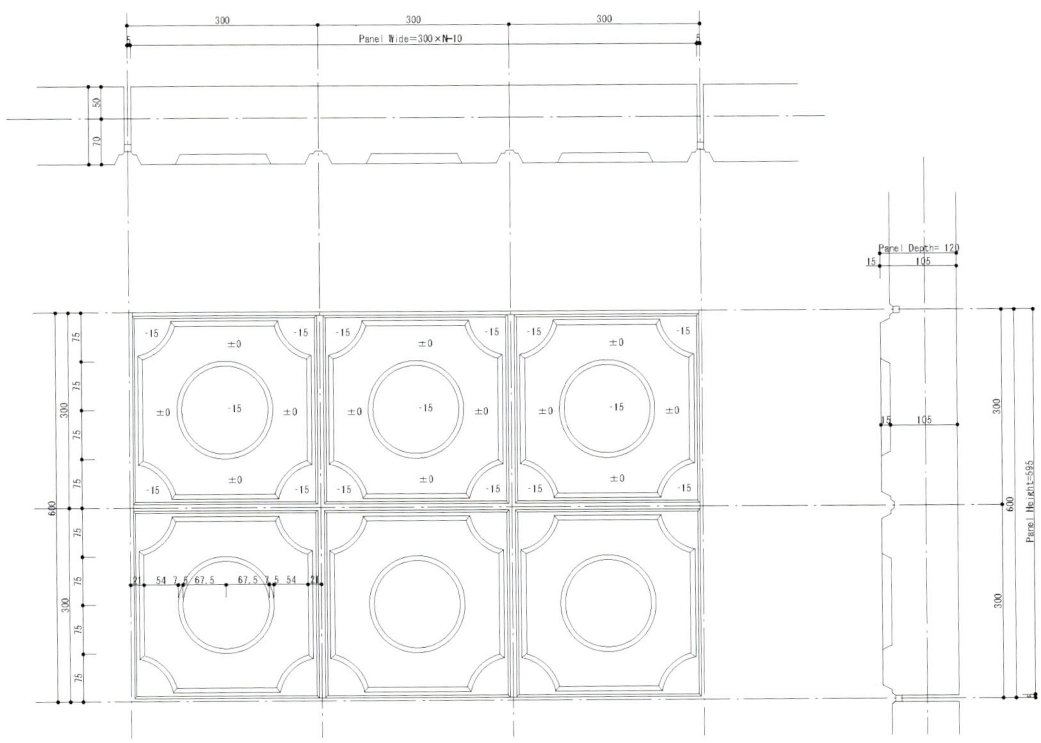

FACADE DETAIL

Interior space that two floor open
View of the inside void space

↑ Protruding space as seen from inside

1 ENTRANCE HALL
2 SHARED CORRIDOR
3 TENANT AREA
4 TERRACE

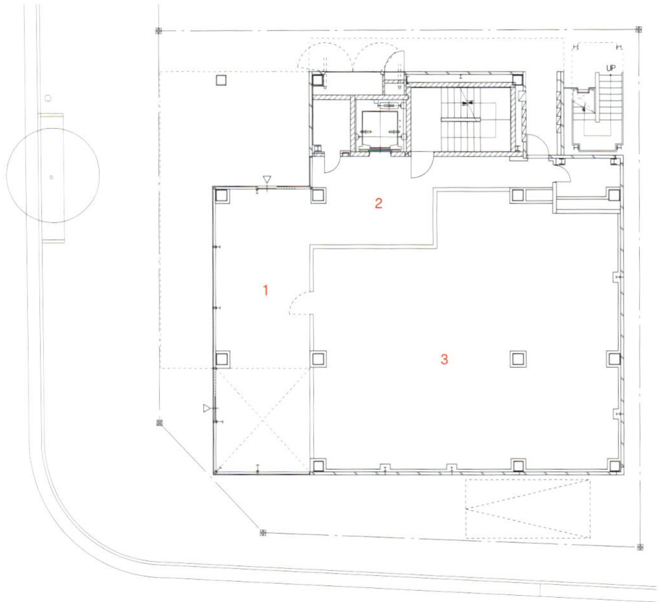

1ST FLOOR PLAN

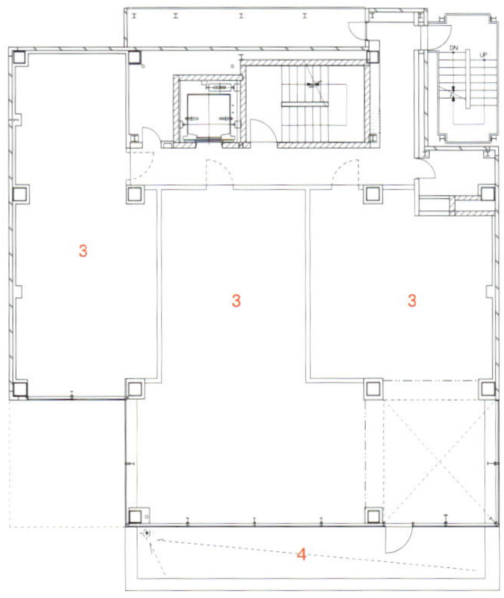

6TH FLOOR PLAN

LK GROUP HEADQUATER

LK 그룹 사옥

It is a building project for LK Group, which is an enterprise to make precision instruments. Only about half of available floor space index was used to for the space with various programs including the building, a distribution center, exhibition and education spaces and an AS center. While the elevation was planned to be closed to cope with inadequate surroundings, all the inner rooms were connected with the courtyard, the balcony and the terrace to secure openness. These spaces that connect interior and exterior have neutral and semi-transparent characteristics. Also it is possible to connect spaces to have expansion and they are transferable. To escape from images of the existing office building, a minimalized form with abundant volume and new materials were planned and it is expected that old buildings would become an icon to give new strength to the street.

정밀 기계를 만드는 기업인 LK Group의 사옥 프로젝트이다. 사용 가능한 용적률의 절반 정도만 사용하여 사무 공간, 물류센터, 전시 및 교육 공간, AS센터 등의 프로그램으로 공간을 구성했다. 열악한 주변 환경에 대응하고자 입면을 폐쇄적인 형태로 계획했지만 내부의 모든 실들을 중정, 발코니, 테라스와 연결하여 개방감을 확보했다. 외부와 내부를 연결하는 이러한 공간들은 중성적이며 반투명한 성격을 가진다. 또한 알맞은 스케일로 공간을 연결하고 확장시켜 상호 전이가 일어나도록 했다. 기존의 사무소 건물 이미지를 탈피하여 풍부한 볼륨감을 가진 미니멀한 형태의 계획과 새로운 재료의 사용으로 오래된 건물들이 늘어선 가로에 새로운 힘을 주는 아이콘과 같은 효과를 기대하였다.

Architect : UAD Architects Group + Mun Jung Il **Location:** 58-2 Yangpyeong-dong, Yeongdeungpo-gu, Seoul, Korea **Site area :** 1,011.90m² **Building area :** 597.44m² **Gross floor area :** 4,205.78m² **Building coverage :** 59.04% **Floor space index :** 259.12% **Construction period :** 2006. 2 ~ 2007. 4 **Photographer :** Park Young chea

설계 : (주)건축사사무소 유에이디 + 문정일 **위치 :** 서울특별시 영등포구 양평동 3가 58-2번지 **대지면적 :** 1,011.90㎡ **건축면적 :** 597.44㎡ **연면적 :** 4,205.78㎡ **건폐율 :** 59.04% **용적률 :** 259.12% **공사기간 :** 2006. 2 ~ 2007. 4 **사진 :** 박영채

RENDERING IMAGE

← Exterior view

1 GROUND PARKING LOT
2 ENTRANCE OF UNDERGROUND PARKING LOT
3 MAIN ENTRANCE
4 LOBBY
5 ELEVATOR HALL
6 OFFICE
7 COURTYARD

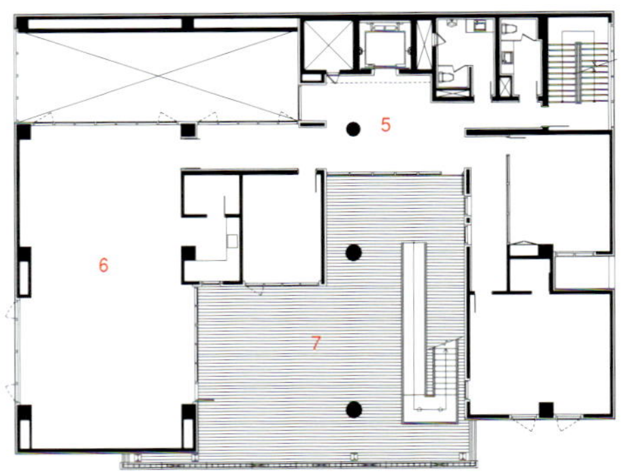

3RD FLOOR PLAN

1ST FOOOR PLAN

⌐ Exterior view ⌐ Exterior view ⌐ Courtyard

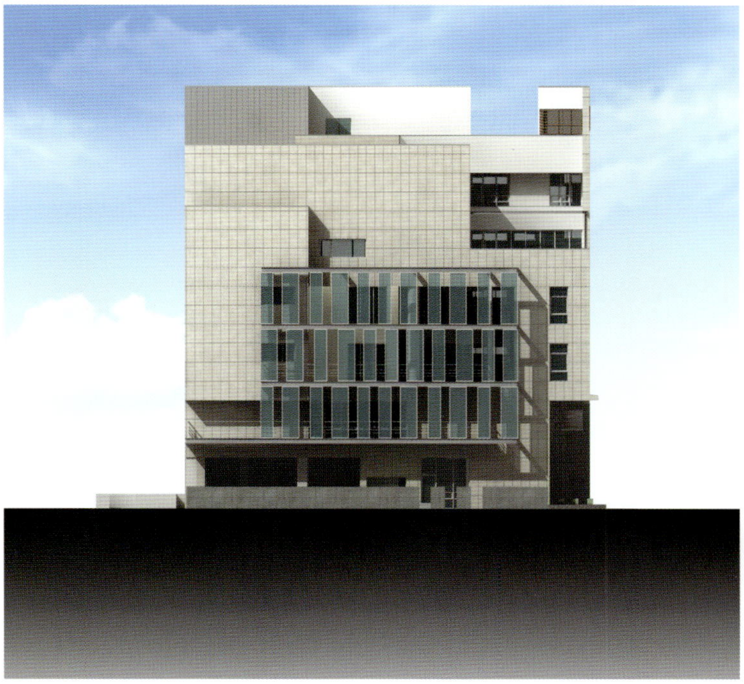

LEFT ELEVATION

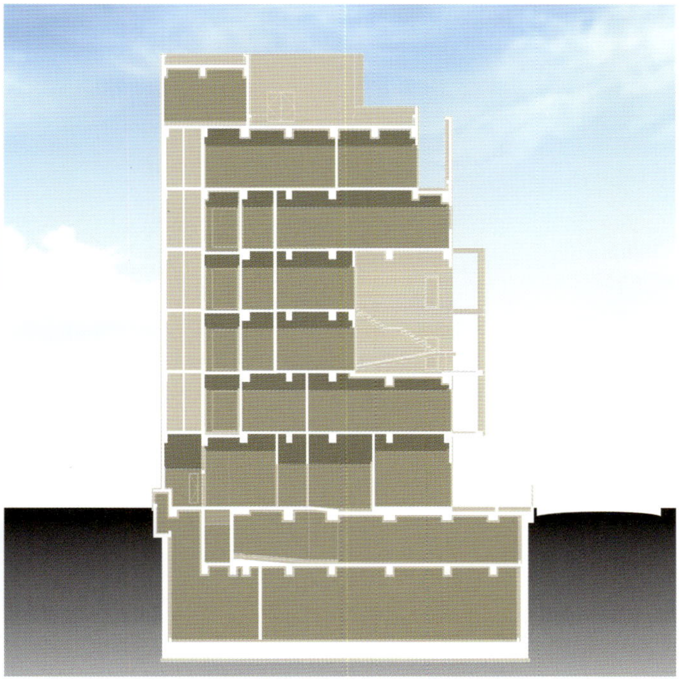

LONGITUDINAL SECTION

↑ Courtyard view from hall ↓ Meeting room

↑ Corridor ↓ Lounge

JUNIOR U STUDIO

주니어유 스튜디오

This project is for a studio building that takes photos of mainly children. Since most photos are taken by using natural light it was necessary to consider light first and external spaces for backgrounds of photos were required. The internal and external spaces were formed by using functions of a camera metaphorically. The shape of the building was conceived to be similar to something like a lego block to create friendliness for children. First of all, a simple basic box to contain functions of a studio was planned. Considering the right to enjoy sunshine and diagonal plane control of street the box was divided into two parts. The northen box was placed on the ground and the other one was lifted up. The transparent volume inserted into each box externally separates boxes and internally connects spaces in those boxes. Each box reflects both internal and external functions and produces open spaces by partially cutting or reducing itself. These external spaces accept light and are used as expanded inner space on each floor. Architectural factors become the minimum background for the studio. The volume was finished with white exposed concrete.

Architect : Archium(Jung Seung Kwon) **Location :** 81-6 Cheongdam-dong, Gangnam-gu, Seoul, Korea **Site area :** 274.9m² **Building area :** 157.51m² **Gross floor area :** 677.74m² **Building coverage :** 57.29% **Floor space index :** 171.04% **Design period :** 2008. 10 ~ 2009. 3 **Photographer :** Park Young Chea

설계 : (주)건축사사무소아르키움(정승권) **위치 :** 서울특별시 강남구 청담동 81-6번지 **대지면적 :** 274.9㎡ **건축면적 :** 157.51㎡ **연면적 :** 677.74㎡ **건폐율 :** 57.29% **용적률 :** 171.04% **설계기간 :** 2008. 10 ~ 2009. 3 **사진 :** 박영채

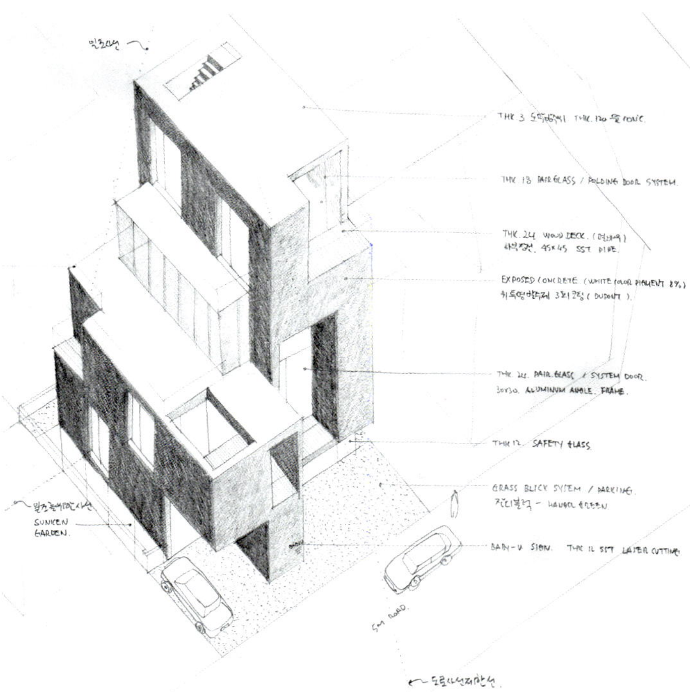

SKETCH

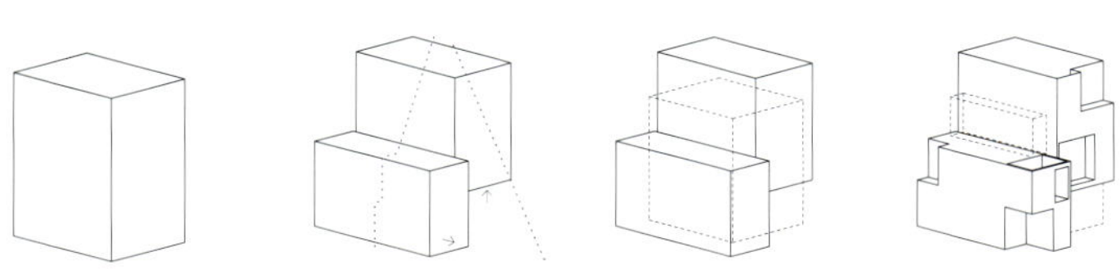

PROCESS DIAGRAM

← Night view of the front right side

↰ Night view ↱ Exterior view

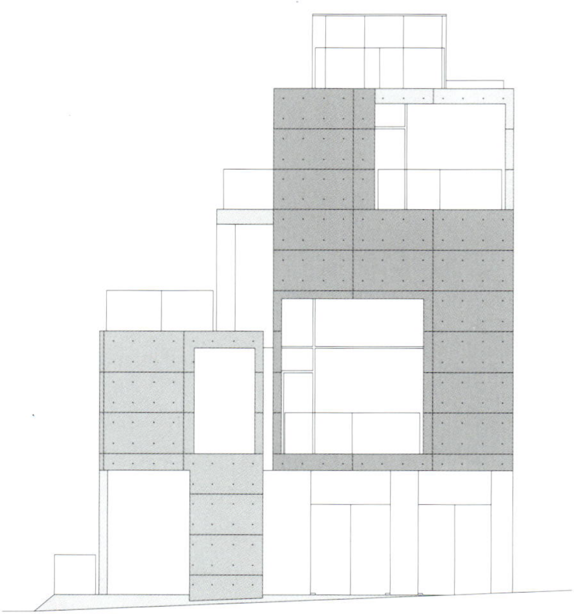

FRONT ELEVATION

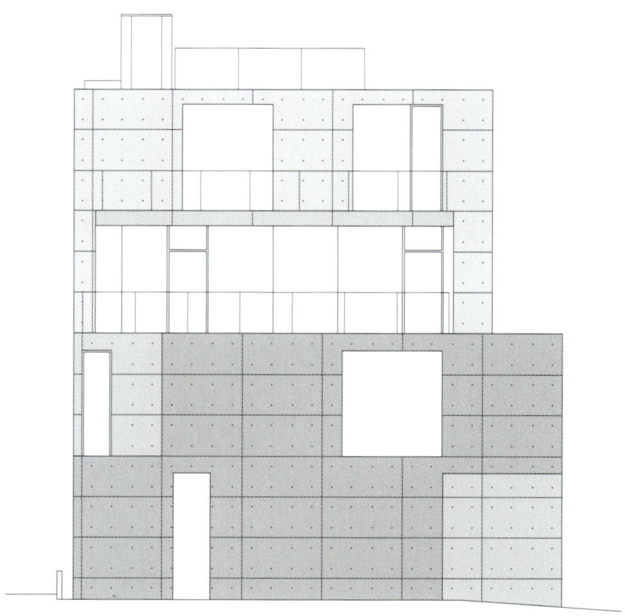

LEFT ELEVATION

↑ Interior view of the 3rd floor ↓ Interior view of the 2nd floor

본 프로젝트는 어린이를 위주로 촬영하는 스튜디오 건물이다. 자연광을 이용해 촬영을 많이 하기 때문에 빛에 대한 고려가 우선되었고, 각 층별로 사진 작업의 배경이 되는 외부 공간이 필요했다. 카메라의 기능을 은유적으로 사용하여 내외부의 공간을 구성했다. 건물의 형태적인 이미지는 레고 블록과 같이 아이들에게 친근감을 줄 수 있는 모양을 떠올렸다. 우선 스튜디오의 기능을 담는 기본적이고 단순한 형태의 상자를 구상했다. 일조권이나 도로의 사선 제한을 고려하여 상자를 둘로 나눈 후 북쪽의 상자는 땅에 내려놓고 다른 하나는 들어올렸다. 각 상자 사이에 삽입된 투명한 볼륨은 외부적으로는 상자들을 분리하고, 내부적으로는 상자 속의 공간을 서로 연결하는 기능을 한다. 각각의 상자들은 내외부의 기능을 반영하여 부분적으로 파내고 덜어내어 열린 공간을 갖도록 한다. 이러한 외부 공간들은 빛을 받아들이며 각 층에서 확장된 실내 공간의 영역으로 사용된다. 건축적 요소들은 스튜디오의 최소한의 배경이 된다. 볼륨은 백색의 노출콘크리트로 마감했다.

↑ Deck of the 4th floor and interior view ↵ Former view ↳ Cafe ↱ View of the 2nd floor from inside to outside

1 DECK
2 STUDIO
3 ENTRANCE DECK
4 GRASS YARD
5 CAFE

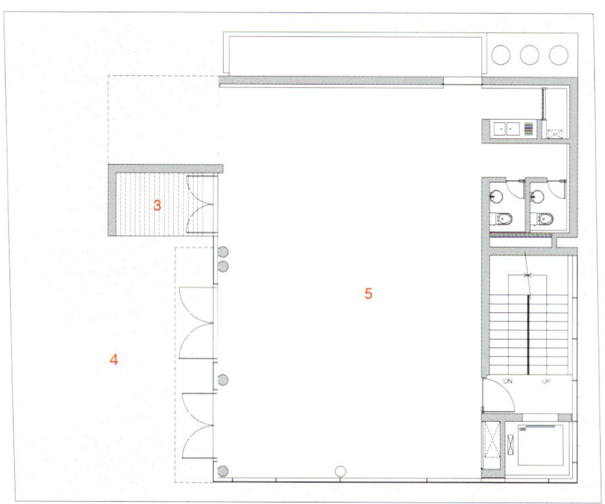

1ST FLOOR PLAN

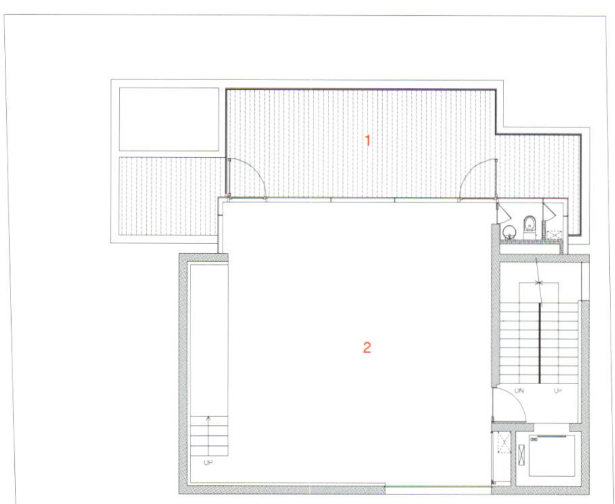

3RD FLOOR PLAN

SEOCHO 1703-9 BUILDING

서초동 1703-9

There was a retaining wall to follow regulation about the right to enjoy sunshine, but it created a physical obstacle at the same time. The long and large retaining wall naturally decided coordinates for the building and the site's width became smaller. To solve deficiency of utility, it was required to reduce unavailable space produced by acute angles at edges more than anything. Diagonal edges were sealed to 'substantialization of the boundary' and reduce unavailable space. As the point of contact was minimized more available floor area was secured. Also from the ground the elevation looks higher due to the thrilling highest point of the curved side and 'uncertainty of the boundary' in some observation locations. These transformations can be defined as internal form exaggeration to use space economically in disadvantageous conditions. Through this, it became to have a dramatic and stable shape. When 'abstract process for a detailed form' is added to overcome observation limitation and undersized mass matters due to a small site in this situation, the overall outline is completed. 'Abstract process for a detailed form' refers to process to visually revise parts that become reduced because of space between a building and its site by using distortion effect from two sides with the same shape and different proportion and sky exposure plane control. What was suggested through the plan that had some distance from characters of the region(judicial town) was not about just solving a problem of the site. It showed that there is possibility of a new design even in disadvantageous conditions when there is active attitude to access a common problem. We hope that this project becomes a model for urban landscape projects in the future.

대지에는 무형의 일조 법규선과 물리적인 방해를 주는 선으로 옹벽이 있었다. 규모가 큰 옹벽은 건물이 들어설 좌표를 자연스럽게 결정하였고, 대지의 폭은 줄어들었다. 유용성 부족을 해결하기 위해 끝으로 갈수록 예각으로 발생하는 불용공간을 줄이는 작업이 우선이었다. 사선의 가장자리에 인장을 가해 긴장감 있는 형태를 만드는 일종의 '경계의 실체화' 조작을 통해 불용공간을 줄이게 된다. 접점의 최소화로 외형적으로는 손실이 있어 보이지만 실제 유용 가능한 바닥면적을 더 확보했다. 또한 지상에서 볼 때 곡면 최고점 부분에서 느껴질 아슬함으로 입면 상승효과 때문에 관찰 위치에 따라 경계점이 모호해지는 '경계의 불명확화'가 연출된다. 이러한 변형은 외부의 불리한 조건에 대해 공간의 낭비를 없애기 위한 일종의 내부적 형태 부풀리기라 정의할 수 있다. 이를 통해 극적이면서도 안정된 형상이 되었다. 여기에 협소한 대지 여건으로 발생하는 투시도적 관찰 한계와 그로 인해 왜소해질 매스를 극복할 수 있는 '구체화된 형상을 추상화' 하는 과정이 더해지면 전체적인 윤곽이 완성된다. '구체화된 형상을 추상화' 한다는 것은 형상은 같지만 비율이 틀린 양 측면을 이용한 왜곡효과로 사선 제한과 함께 대지 내 주요 제한 사항인 건물과 대지의 이격으로 인해 축소된 부분을 시각적으로 보정한다는 의미가 있다. 지역적 특성(법조타운)과 다소 거리가 있는 형태의 제안을 통해 의도한 것은 단순히 한 대지만의 문제를 해결하는 차원이 아닌 공통적인 문제에 대해 적극적인 방법으로 접근을 한다면 불리한 조건에서도 새로운 디자인 가능성이 있음을 보여주는 것이다. 앞으로의 도시의 경관 변화에 대해 이 프로젝트가 하나의 전형으로 남길 바란다.

← Exterior view

Architect : SEEGAN ARCHITECTS(Park You Jin) **Location :** 1703-9 Seocho-dong, Seocho-gu, Seoul, Korea **Site area :** 350.90m² **Building area :** 178.15m² **Gross floor area :** 802.75m² **Building coverage :** 50.77% **Floor space index :** 186.42% **Construction period :** 2007. 6 ~ 2008. 3 **Photos offer :** SEEGAN ARCHITECTS

설계 : (주)건축사사무소시간(박유진) **위치 :** 서울특별시 서초구 서초동 1703-9번지 **대지면적 :** 350.90㎡ **건축면적 :** 178.15㎡ **연면적 :** 802.75㎡ **건폐율 :** 50.77% **용적률 :** 186.42% **공사기간 :** 2007. 6 ~ 2008. 3 **사진 :** (주)건축사사무소시간 제공

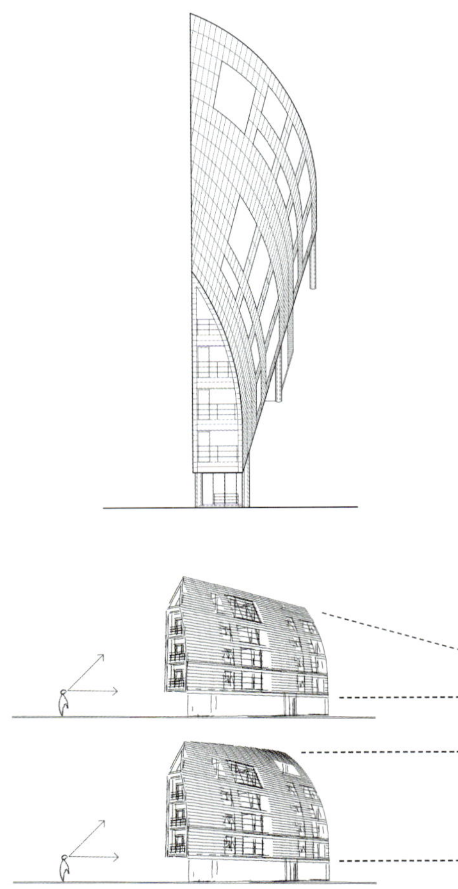

DIAGRAM

↰ Night view from East
← Exterior view of upper part

↑ Interior view

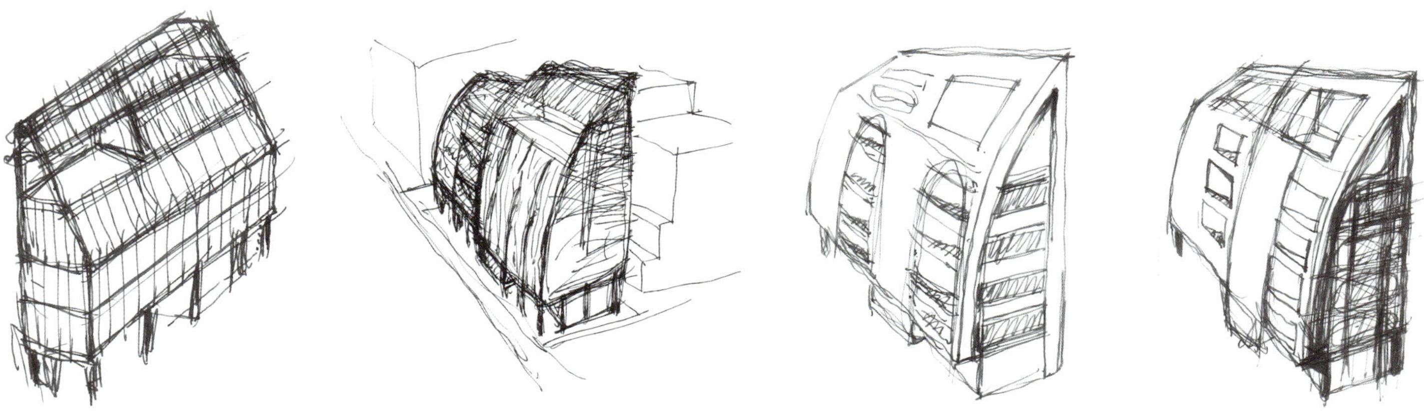

VOLUME STUDY SKETCH

1 PARKING LOT
2 HALL
3 NEIGHBORHOOD FACILITY
4 TERRACE

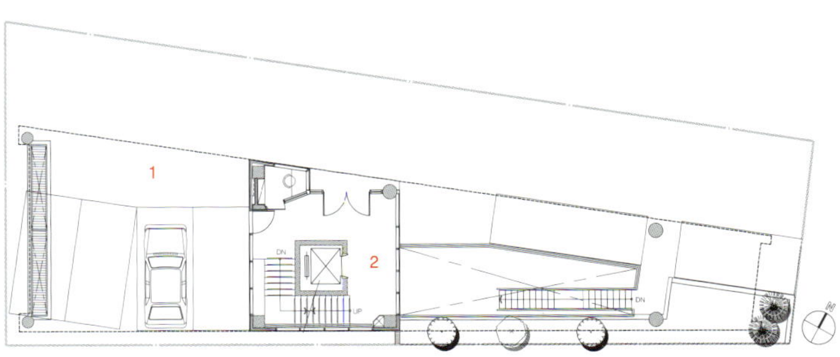

1ST FLOOR PLAN

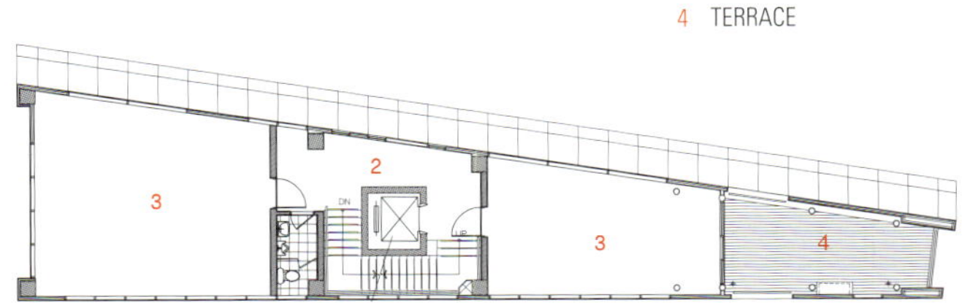

5TH FLOOR PLAN

← View from East ↰ Interior view of the 5th floor ↱ Sunken space

NORTH ELEVATION

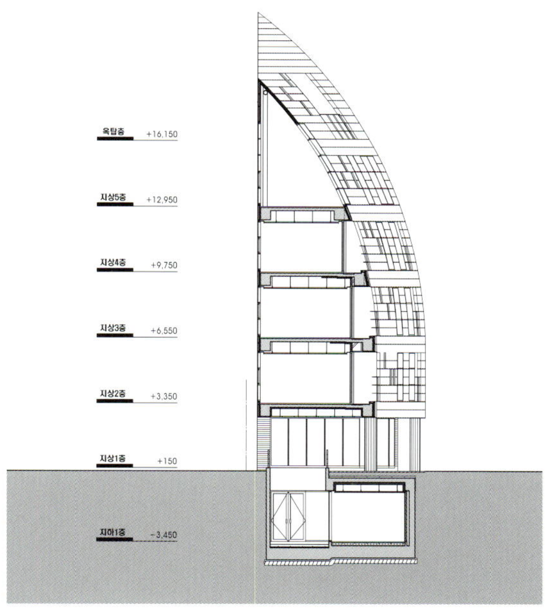

LONGITUDINAL SECTION

MONACO HOUSE

모나코 하우스

This is a new 4 storey building located in a largely pedestrian lane(Ridgway place) at the East end of Melbourne's CBD. Dominant in the lane is the historic Melbourne Club wall and the gigantic plane trees emanating from the Melbourne Club Garden which arch over the diminutive lane. The site with dimensions of 6.1m in width and 17m in depth and a foot print of 102.5m² is a postage stamp. Our brief was to provide a ground level entry and cafe, followed by two levels of office tailored for the Proprietors Investment and Philanthropic Organisation. The top level contains a small reception area primarily for official functions associated with the client's role as Honorary Consular of Monaco. We wanted the building to be both abstract and awash with imagery. We looked at the plane trees, the gothic, surrealism, the heraldic, deco and the Prague cubists. We wanted the building to be above all else something that amplified its miniature urban grain and enriched the pedestrian experience of the city. Within the office large apertures to the West are well shaded by deep balconies and the adjoining plane trees. The workspace has good natural light and cross ventilation. Exposed windows have electronically controlled external blinds. Outdoor balconies provide areas of release from the office desk. The 'green roofscape' is a similar space but also adds additional insulation to the upper floor. From a technical point of view the complex geometry and use of such materials as Glass Reinforced Cement and its structural support demanded sophisticated use of 3D software and integration of those techniques with the off site production.

Architect : McBride Charles Ryan(Robert McBride, Debbie-Lyn Ryan) **Location :** Melbourne, Australia
Building area : 400m² **Completion :** 2007 **Photographer :** Trevor Mein

설계 : 맥브라이드 찰스 라이언(로버트 맥브라이드, 데비-린 라이언) **위치 :** 오스트레일리아 멜버른 **건축면적 :** 400㎡
완공 : 2007 **사진 :** 트레버 마인

← Exterior view ⌝ Meeting area ⌜ Workspace

← Folding surfaces create a play of light and shadow ↰ View of the corner structure ↱ Detail of the folding surfaces ↲ View from North side way

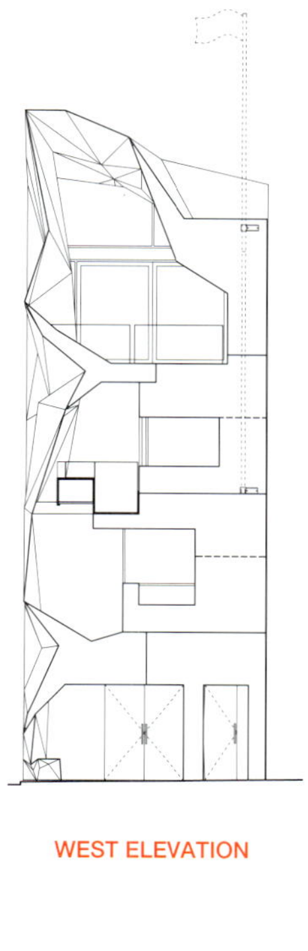

WEST ELEVATION

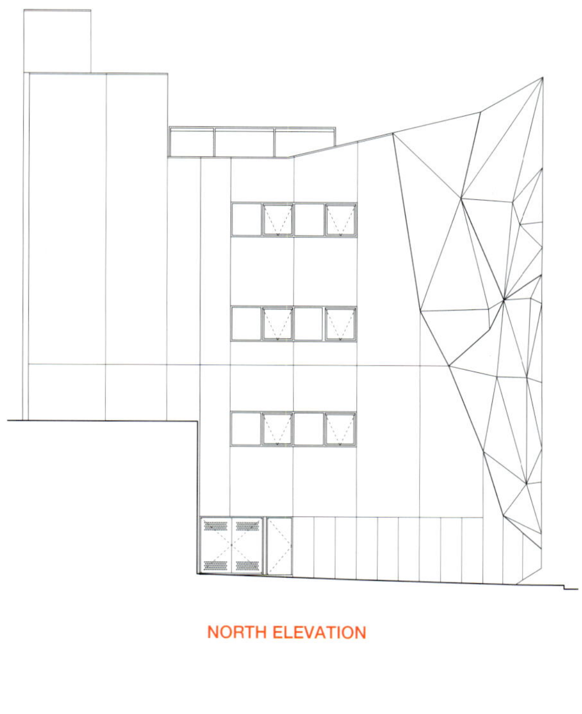

NORTH ELEVATION

이 프로젝트는 4층의 신축 건물로, 멜버른 중심가 동쪽 끝에 주로 보행자가 오가는 좁은 길에 자리 잡고 있다. 주변에서 가장 눈에 띄는 것은 역사가 오랜 멜버른 클럽 벽과 플라타너스 나무들이다. 건축주는 1층에 입구와 카페를 만들고, 2개 층의 사무실을 원했다. 꼭대기 층에는 주로 건축주와 연관된 모나코 명예영사관의 공식적 기능을 위한 소규모의 리셉션 구역이 구성된다. 우리는 건물에 추상적이고도 갖가지의 이미지를 지니도록 하고 싶었다. 무엇보다도 플라타너스 나무, 고딕, 초현실주의, 문장학, 아르데코, 프라하의 입체파 등을 살피며, 건물이 축소된 도시조직을 증폭시켜 도시 내의 보행 경험을 풍부하게 하는 존재로 자리 잡게 하고 싶었다. 사무실 내에 커다랗게 뚫린 서향의 개구부는 인접한 플라타너스 나무와 깊은 발코니 덕분에 그늘이 잘 드리운다. 작업공간은 자연광이 잘 들며 양쪽으로 통풍이 잘 이루어진다. 노출된 창에는 전자식으로 제어되는 외부 블라인드가 설치됐다. 사무실 책상으로부터 벗어난 옥외 발코니는 휴식 영역이 된다. '녹색의 지붕 경관' 역시 이와 비슷한 공간으로 위층 상부에 또 한 켜의 단열층을 형성한다. 기술적으로 복잡한 기하학 형태, 유리섬유 강화 시멘트 등의 소재 사용 그리고 구조체 구현을 위해 정교한 3D 소프트웨어를 이용할 뿐 아니라 사전 제작 기법을 통합해야만 했다.

← Internally, the walls flow on to become the upholstered ceiling ↑ Interior view of the 4th floor

1 CAFE
2 FOYER
3 ENTERTAINING

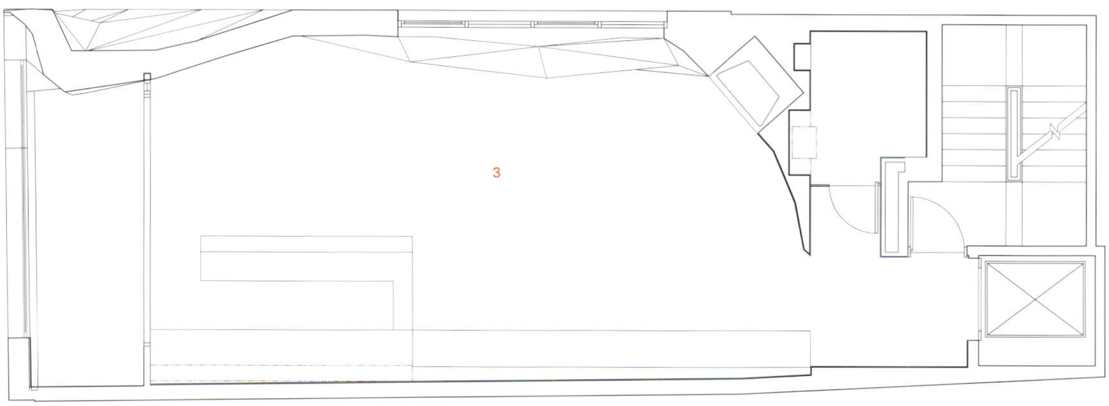

LEVEL+3 FLOOR PLAN

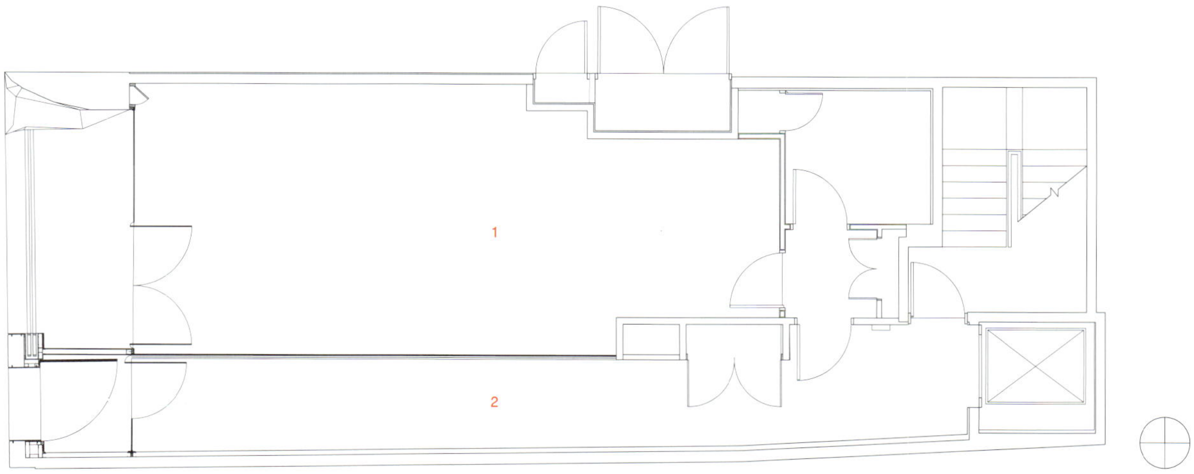

GROUND FLOOR PLAN

MOKUZAI KAIKAN

목재 협회관

A semi - outdoor space called engawa, existing in traditional Japanese homes, allows a natural breeze to enter the shelter while shutting out strong sunlight, achieving a comfortable indoor environment in the midst of harsh heat and humidity. From the traditional engawa, a new space configuration was realized. In this configuration, elements of an office, such as its structure, circulation system, and facilities, are reconstructed in a semi-outdoor environment. This configuration strives to connect the outdoors with the universal office, while simultaneously acting as a buffer against the harsh side of the external environment. Sets of terraces come together to make a porous skin. There is an aspect of Japanese spatial composition that makes use of layering and preserves spatial continuity. Instead of building a wall to divide spaces, fixtures are installed to give an abstracted impression of the adjacent space. In this way, spaces flow from one into other. This building achieves this spatial richness. Rooms, walls, and the roof are dispersedly located, giving movement and flow to the scene, while a sense of the next space is subtly delivered, via the sights that pass through or the light that enters into an area.

전통 일본 가옥의 '엔가와'(반 외부 공간)를 이용하여 새로운 공간 배치를 했다. 사무 공간 요소, 구조, 동선 체계와 시설 등을 반외부 공간을 통해 재구성했다. 이러한 배치 구조는 외부 공간을 내부 사무 공간과 소통시켜 풍부한 내부 환경을 제공하고, 외부 환경의 불필요한 요소를 완충하는 역할을 한다. 반외부 공간 구조체인 테라스는 마치 볼륨의 타공 외피로 보인다. 공간을 겹쳐 연속성을 부여하는 일본식 공간 구성으로 공간을 구분 짓는 벽 대신 인접한 공간에 대한 추상적 표현으로 가벽을 설치하였다. 이러한 방식을 통해 공간은 서로 소통하며 풍부해진다. 벽체와 지붕은 무작위로 배치되어 움직임과 흐름을 생성하고, 공간을 통과하는 광경과 공간 내로 들어오는 빛을 통해 근접한 공간의 존재감을 느낄 수 있다.

Architect : NIKKEN SEKKEI(Tomohiko Yamanashi + Takeyuki Katsuya) **Location :** Tokyo, Japan **Site area :** 1,652.90m² **Building area :** 1,011.26m² **Gross floor area :** 7,582.09m² **Construction period :** 2007. 11 ~ 2009. 6 **Photographer :** Nacasa & Partners. Inc, Harunori Noda

설계 : 닛켄 세케이(토모히코 야마나시 + 타케유키 카츠야) **위치 :** 일본 도쿄 **대지면적 :** 1,652.90㎡ **건축면적 :** 1,011.26㎡ **연면적 :** 7,582.09㎡ **공사기간 :** 2007. 11 ~ 2009. 6 **사진 :** 나카사 & 파트너스, 하루노리 노다

← View of the main facade ↰ The part of view of the facade ↱ Concept furniture

← The part of view of the facade ↑ General view

1 TERRACE
2 OFFICE
3 WOOD LOBBY
4 MEETING SPACE
5 GALLERY
6 GARDEN
7 PARKING LOT
8 COMMUNITY PLAZA

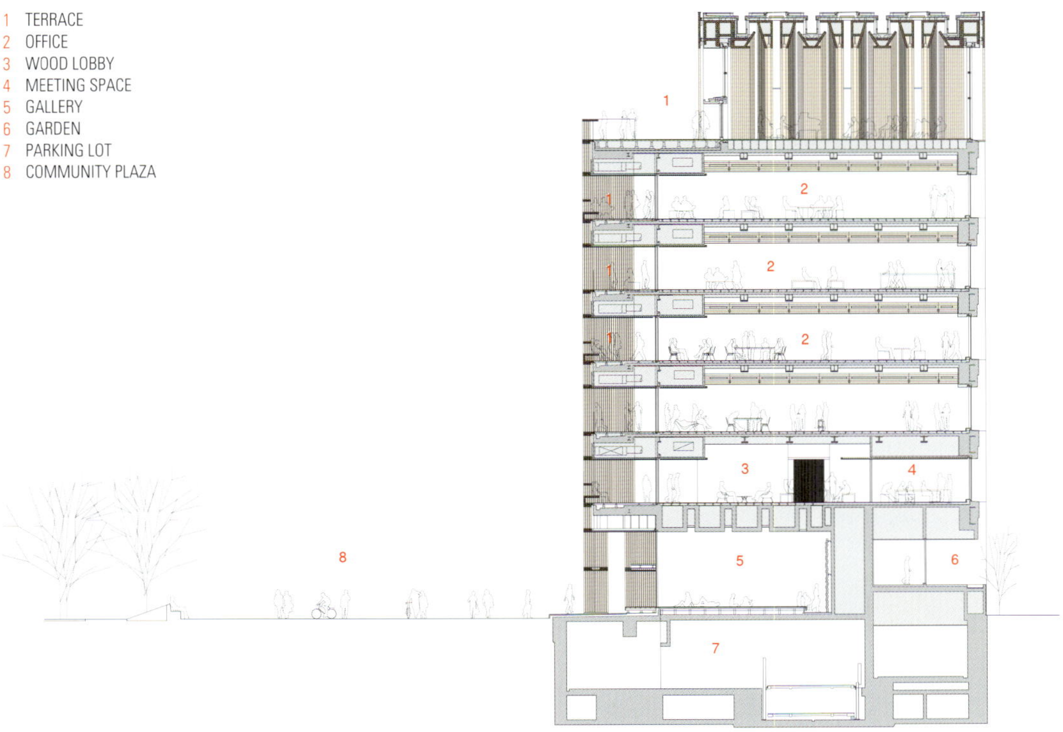

EAST-WEST SECTION

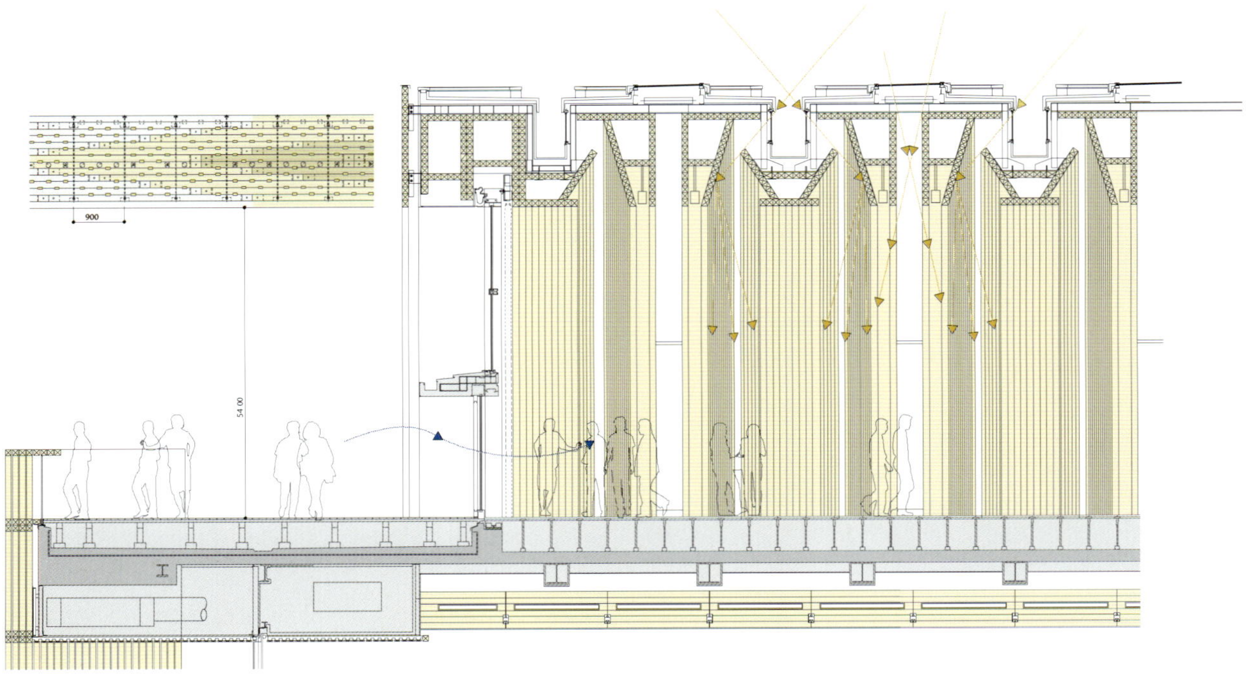

SECTION DETAIL

↰ Interior view of the 7th floor ↑ Interior view of the 7th floor ↲ Interior detail ↳ Interior detail

↑↓ Gallery

↑ Entrance hall ↲ Terrace ↳ Interior view

1 TERRACE
2 GALLERY
3 ENTRANCE HALL
4 WOOD LOBBY

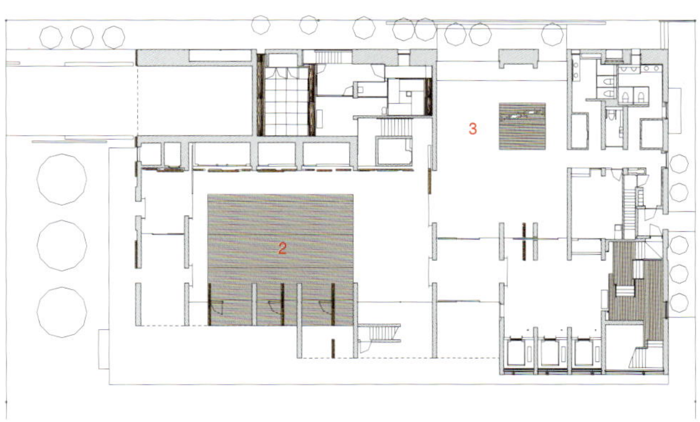

1ST FLOOR PLAN

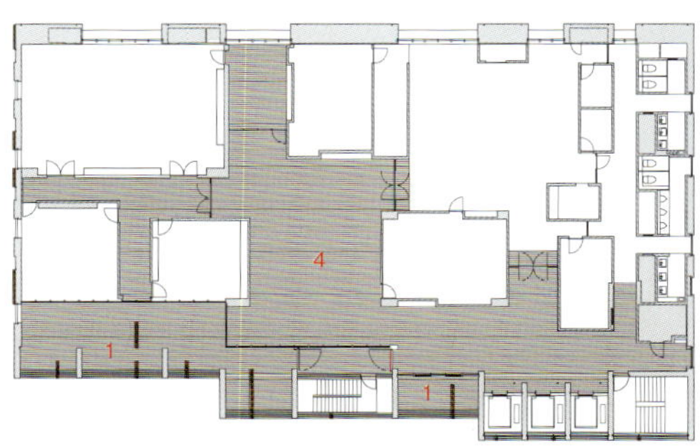

2ND FLOOR PLAN

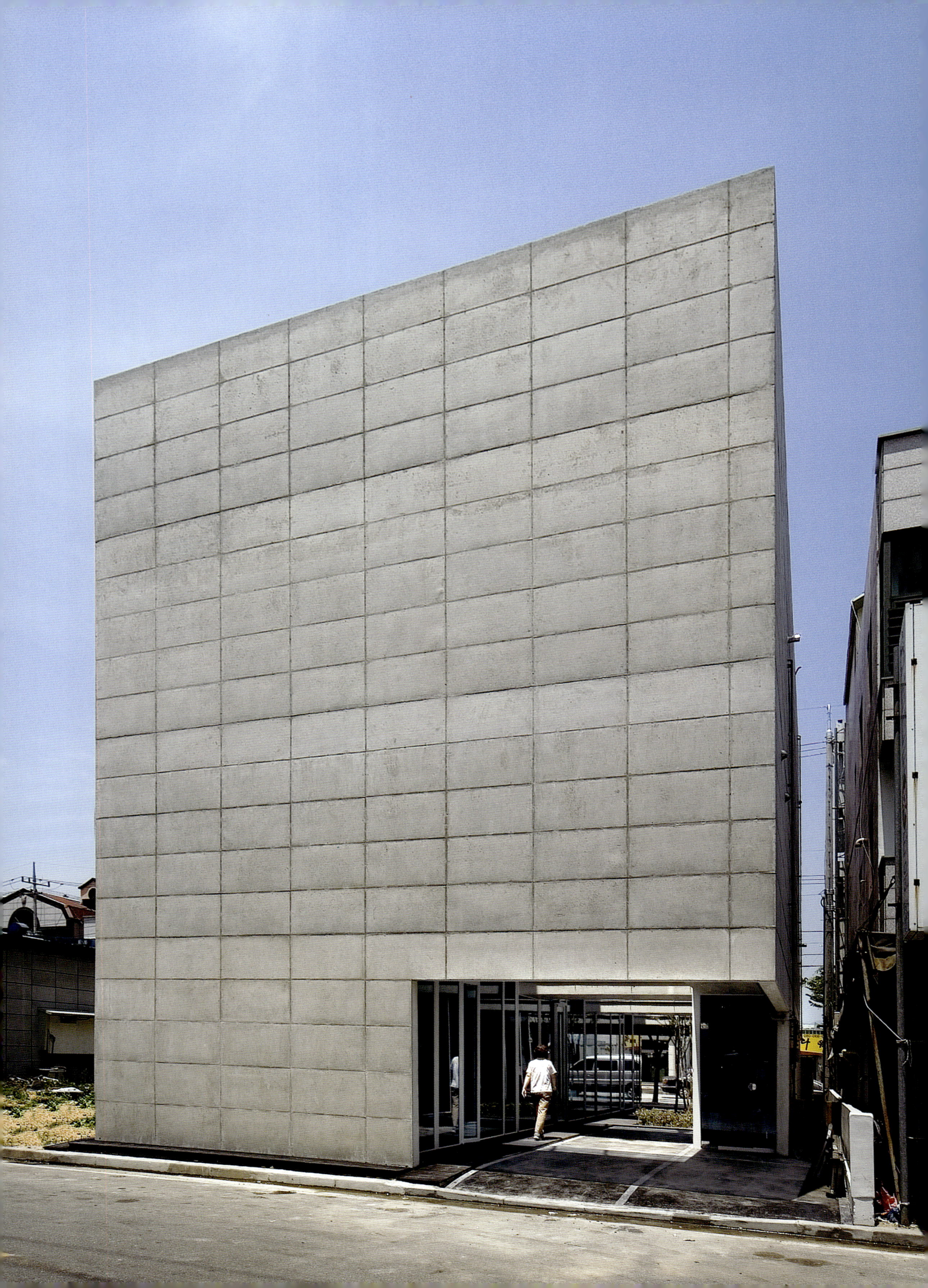

INVOID

순천 최종영 신경정신외과

I intended to create an inside path which connects streets front and rear. The house is laid on a road when the road is opened and the earth is emptied. It does not need to make a detour if the road is always opened. The purpose of this is not for just a justification as convenience of residents and communication of space, but for utility of the owner. The house owned by a doctor should not be overlapped with visit from patients. It would be more general when open the road to keep composure than dividing the entrances and blocking them. I arranged the main space on the right side in a long shape and stairs on the opposite side, so they are continued to the path. For the path which is linked like an overpass is laid with space, vertical opening space forms overlapped shape over the path. The simple cube shape is enough for the form of space-horizontal and vertical void is crossing.

앞길과 뒷길을 이어주는 안길을 만들기로 했다. 길을 열어 땅을 비우면 집은 길 위에 놓이게 된다. 언제나 열려 있을 길을 만들면 큰 길을 돌아가지 않아도 된다. 주민의 편리함이나 공간적 소통과 같은 명분이 아니라 주인에게 도움을 주려는 것이다. 의사인 건축주의 집은 환자들의 방문과 겹치지 않아야 한다. 입구를 나누어 막는 것보다 길을 열어 여유를 만들면 넉넉할 것이다. 주 공간을 오른쪽에 길게 두고 왼쪽에 계단을 두어 길 위로 이어지게 만들었다. 육교처럼 연결된 통로는 사이를 두고 놓이므로 수직의 열린 공간이 길 위에 겹쳐진 모양이 되었다. 수평과 수직의 보이드가 교차하는 공간의 형태는 단순한 상자로 충분하다.

Architect : Kim In Cheurl(Chungang University) + Archium (Jung Seung Kweon) **Location :** 1495-3 Yeonhyang-dong, Suncheon, Korea **Site area :** 230m² **Building area :** 137.730m² **Gross floor area :** 367.33m² **Building coverage :** 59.88% **Floor space index :** 159.71% **Construction period :** 2005. 10 ~ 2006. 6 **Photos offer :** Archium

설계 : 김인철(중앙대학교) + (주)건축사사무소아르키움(정승권) **위치 :** 전라남도 순천시 연향동 1495-3 **대지면적 :** 230㎡ **건축면적 :** 137.730㎡ **연면적 :** 367.33㎡ **건폐율 :** 59.88% **용적률 :** 159.71% **공사기간 :** 2005. 10 ~ 2006. 6 **사진 :** (주)건축사사무소아르키움 제공

← View of the facade ↑ Entrance

↑ Entrance → External void space / Bridges

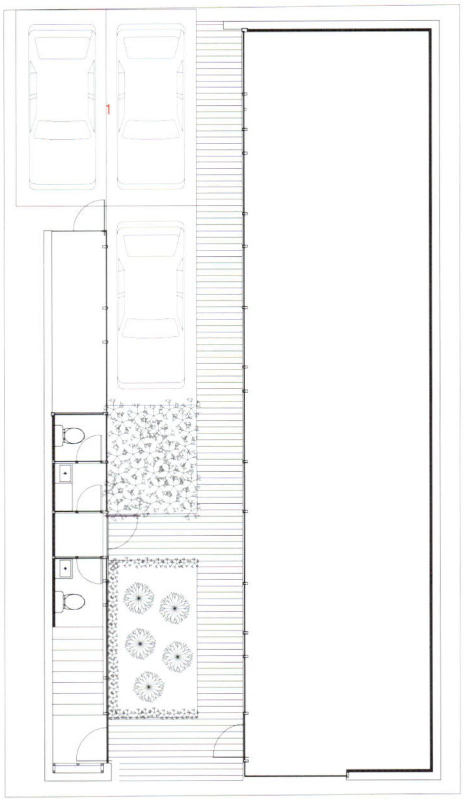

1ST FLOOR PLAN

2ND FLOOR PLAN

1 PARKING LOT
2 VOID
3 BRIDGE

← External void space ↑ View of the 2nd floor ↓ Stairs with courtyard

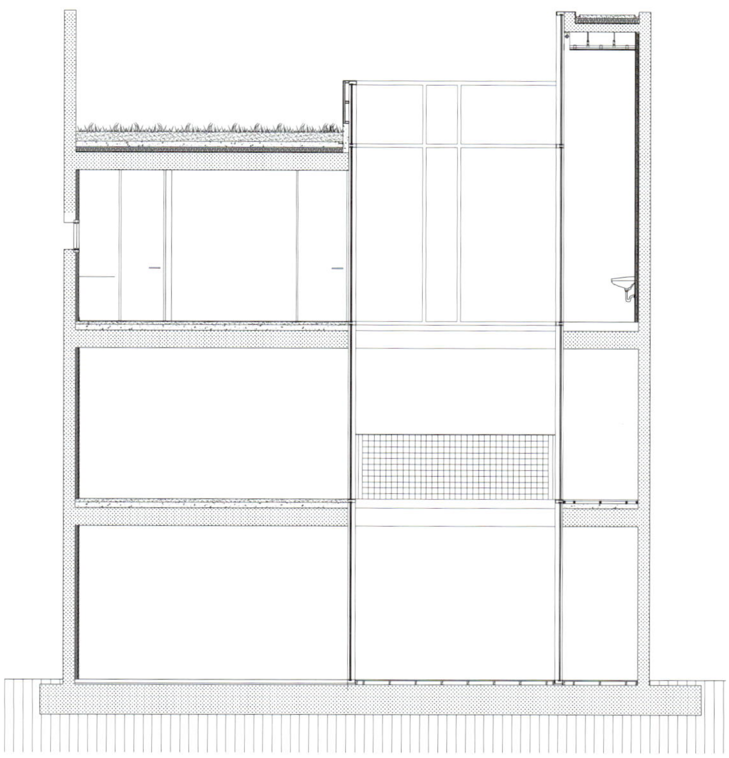

SECTION

…Beyond Creative Potential

DEVOID

두김스튜디오

Space exists in a state of nonexistence. So it's like extorting a confession showing the substance of space. The boundary of existence and nonexistence should be defined first in order to recognize nonexistence. Things not being sensed are appeared to be nonexistence. One's sense can't be presented to be real image while the existence of 0(zero) can be showed in mathematics. It needs to give real evidence rather than showing an impression. Or an experiential evidence can lead to confess. It is specific enough to be conscious of its existence although density and quantity of it was in the state of sense. Compression of sense makes nonexistence conscious. I intended to show the substance of nonexistence by compressing it and set aside all common-sense considerations. Devoid means 'remove the void'. Space which has little openings except for ones only made by functional reason seems to be very closed. But the weight, viscosity, and texture of the space is not exclusive because the essence shown above cannot be confined to one place.

공간은 없음의 상태로 존재한다. 그럼에도 공간의 실체를 증명하는 것은 자백을 강요하는 것과 같다. 없는 것을 있다고 하려면 우선 '있고 없음' 의 한계를 규명해야 한다. 감각되지 않는 것은 없음으로 인식된다. 수학의 논리로는 0의 존재를 증명할 수 있지만 감각은 가상의 것을 실상으로 제시할 능력이 없다. 심증으로 주장하는 것보다 물증을 제시하는 것이 확실할지 모른다. 아니면 경험의 증거를 내밀어 자백을 얻을 수도 있을 것이다. 없음의 절대적인 존재감은 감각의 압축에 의해서 만들어진다. 상식적인 없음을 압축하여 없음의 실체를 증명하려고 했다. 상식의 틀을 벗어나는 상황을 만들기 위해 상식이 배제된 상태를 만들었다. 상식적인 보이드가 없어지면 개념적인 보이드의 실제가 나타날 것이다. 그래서 최소한의 기능적인 개구부만 갖는 공간은 매우 폐쇄적이다. 그러나 형성된 공간의 무게와 점성과 질감은 전혀 가두어져 있지 않다. 찾아낸 공간의 본질은 가두어지는 것이 아니기 때문이다.

Architect : Kim In Cheurl(Chungang University) + Archium (Jung Seung Kwon) **Location :** 401-18 Seogyo-dong, Mapo-gu, Seoul, Korea **Site area :** 342.6m² **Building area :** 181.6m² **Gross floor area :** 1,122.0m² **Building coverage :** 53.02% **Floor space index :** 218.00% **Photos offer :** Archium

설계 : 김인철(중앙대학교) + (주)건축사사무소아르키움(정승권) **위치 :** 서울특별시 마포구 서교동 401-18 **대지면적 :** 342.6㎡ **건축면적 :** 181.6㎡ **연면적 :** 1,122.0㎡ **건폐율 :** 53.02% **용적률 :** 218.00% **사진 :** (주)건축사사무소아르키움 제공

← View from Southeast ↰ Inner void space ↱ Interior view

← Detail view of Northeast facade ↱ Entrance ↓ Inner void space → Inner void space

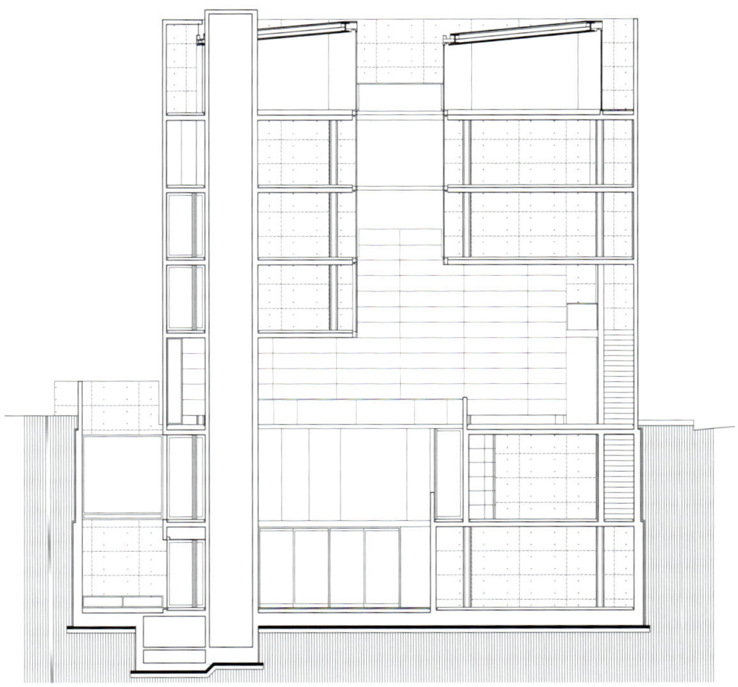

SECTION

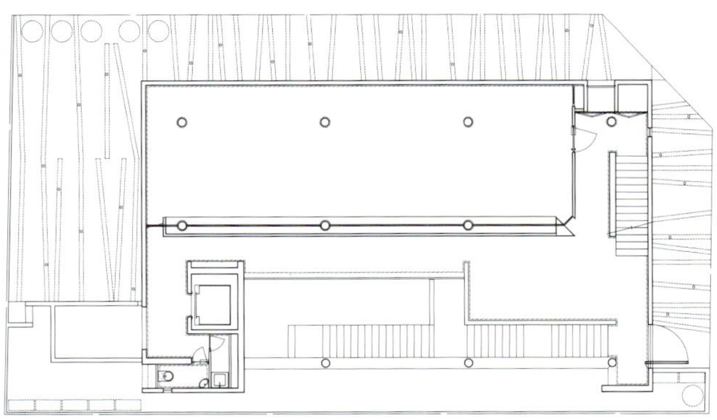

1ST FLOOR PLAN

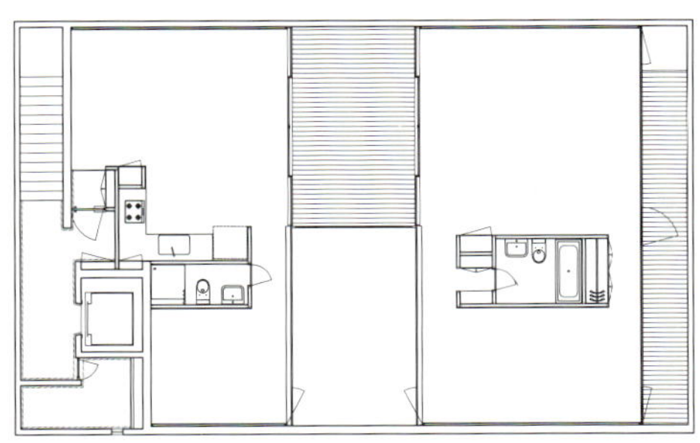

5TH FLOOR PLAN

DOGDECO

도그데코

This project is for a commercial and residential complex on the site that has been informed to be commercial development area. Considering the nearby park and natural contexts including big metasequoia trees, the external was finished with woods with fire - resistant qualities. The client, owner of a pet goods company, wanted to have a store with a residential space for his family. For the 1st level steel concrete structure and frame structure were established to emphasize roads on both sides and improve spatial perception while a glass case surrounded the exposed structure. Behind the building a piloti parking lot and an entrance earth-floored room for dwelling were planned. Skylights above the stairs were planned to induce light into the building. The 2nd level was constructed to contain warmth of woods unlike the 1st level. The kitchen was planned to have level difference in the Japanese Kotatsu style (people can put their feet in the space created under the floor). The living room was made based on life style of sitting on the floor and was connected with the roof balcony making no space for the middle section to the exterior. DOGDECO is a space that combines public section and private section. The way life gets closer to work and work gets close to life shows the client's view of life and it is also expressed through the architecture.

본 프로젝트는 상점 겸용 주택으로 대지는 상업 개발 예정지로 알려진 구역이다. 공원과 인접하고, 커다란 메타세콰이아가 자라는 주변의 자연 컨텍스트를 고려하여 외부는 내화력을 가진 목재로 마감했다. 건축주는 애견 용품 회사의 경영자로 가족과 함께 생활할 주택을 겸한 노면 상점을 원했다. 1층은 양방향의 도로를 살리고 인지성을 높이도록 철근콘크리트조에 라멘조를 살려 노출 구조체를 유리 케이스로 감싸는 형상의 공간으로 계획했다. 건물 뒤편으로 필로티 주차장과 함께 주거용 현관 토방을 마련했다. 계단 상부에 천창을 계획하여 빛이 들어올 수 있게 했다. 2층 공간은 1층의 이미지와 다르게 목재의 온기를 느낄 수 있게 했다. 주방은 일본의 코타츠 형식(바닥, 마루에 공간을 내어 발을 넣을 수 있는 형식) 카운터로 단차를 두어 계획했다. 거실도 좌식생활 공간으로 계획했으며, 루프 발코니와 연속되도록 하여 외부와 내부의 중간영역을 만들지 않았다. 도그데코는 공적 영역과 사적인 영역이 융합된 공간이다. 생활이 일에 가까워지고 일이 생활에 가까워지는 생활 방식은 건축주의 인생관을 나타내며, 건축에서도 동일한 개념을 표현했다.

Architect : APOLLO Architects & Associates(Satoshi Kurosaki) **Location :** Musashino-si, Tokyo, Japan **Site area :** 76.27m² **Building area :** 52.20m² **Gross floor area :** 91.32m² **Building coverage :** 68.44% **Floor space index :** 119.73% **Construction period :** 2008. 5 ~ 2009. 1 **Photographer :** Masao Nishikawa

설계 : 아폴로 아키텍츠 & 어소시에이츠(사토시 쿠로사키) **위치** : 도쿄 무사시노 **대지면적** : 76.27㎡ **건축면적** : 52.20㎡ **연면적** : 91.32㎡ **건폐율** : 68.44% **용적률** : 119.73% **공사기간** : 2008. 5 ~ 2009. 1 **사진** : 마사오 니시카와

← View from Northwest └ View of the kitchen from living room ┘ Kitchen with skylight

← View from North ↑ Shop on the 1st floor

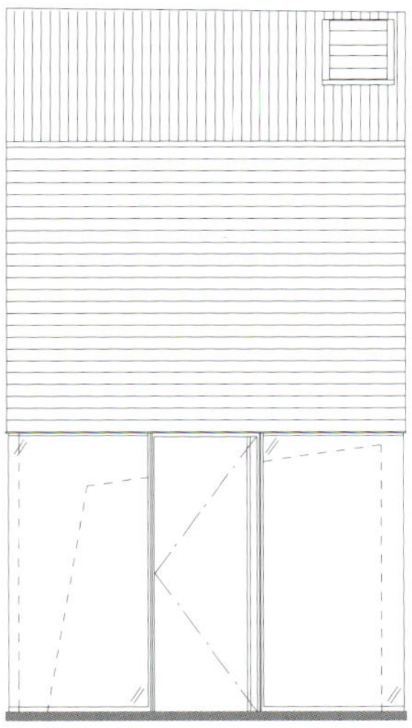

NORTH ELEVATION

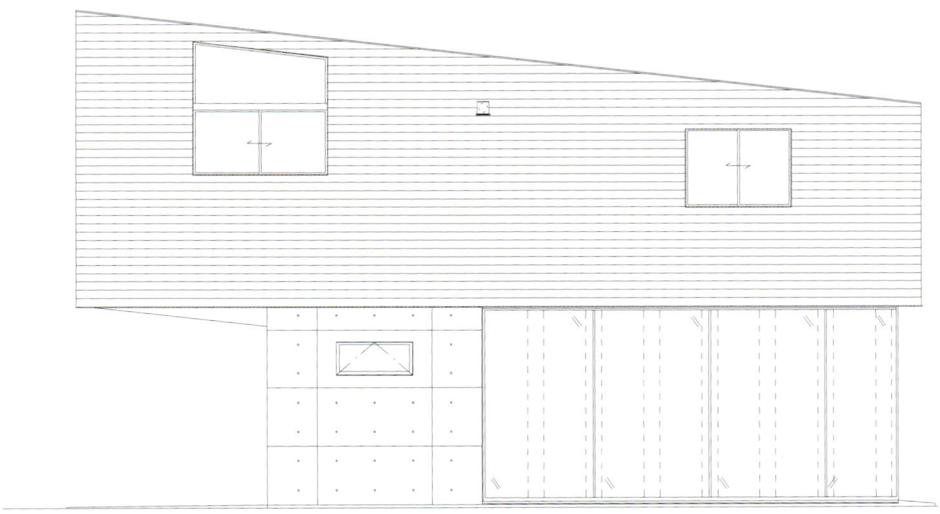

WEST ELEVATION

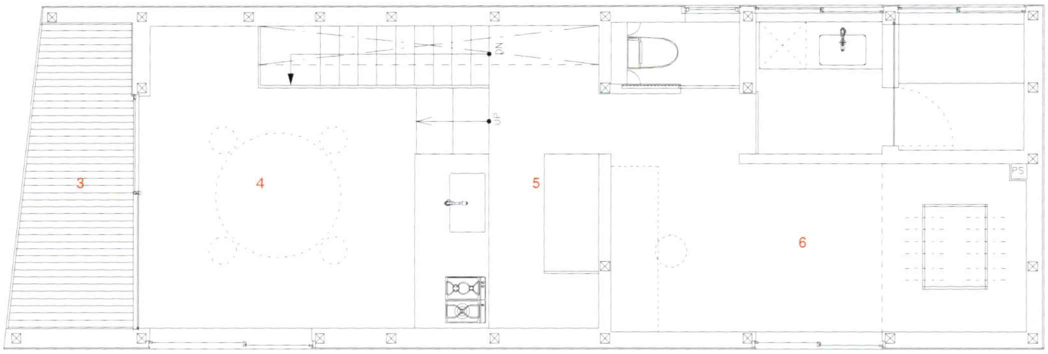

2ND FLOOR PLAN

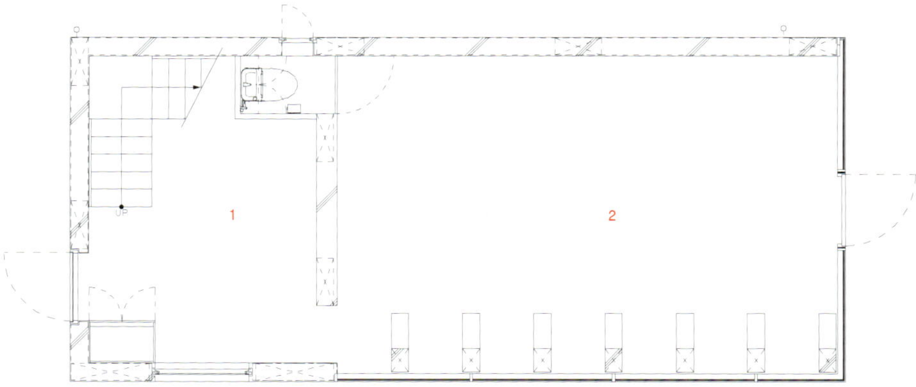

1ST FLOOR PLAN

1 STORAGE
2 SHOP
3 BALCONY
4 LIVING / DINING ROOM
5 KITCHEN
6 BEDROOM

← RC structure ↑ Interior view, 1F ↓ interior view, 1F

← View of the skylight from stairs → Stairs

SECTION

7 30 - 9
16 - 18

SWITCH

스위치

The project is the reconstruction of a 85-years old Unagi(eel)-specialized restaurant located in Ohoi, Shinagawa. As the client wants to welcome guests in a brand-new space in commemoration of the 100th anniversary, we planned the modern concrete buildings within the downscaled site due to expansion of the adjacent road. The site is wide in the front while it tapers in the rear. Reinforcing the advantage of the area of the front, we reduced the open space and designed the lodge wall made of black cedar plywood to express the traditional impact. The living space on the 2nd floor creates a contrast in the exposed concrete. The 4.5m cantilevered balcony is in the shape of a ship.

프로젝트는 시나가와 오오이에 위치한 85년의 전통적인 우나기(장어) 레스토랑의 재건축 작업이다. 도로 확장으로 어쩔 수 없이 축소된 대지에 창업 100주년을 새로운 공간에서 맞이하겠다는 건축주의 의지로 콘크리트의 모던한 형태의 근린생활시설을 계획하게 되었다. 대지는 정면이 넓고 뒤로 갈수록 폭이 좁아진다. 정면의 면적이 큰 이점을 살려서, 1층은 열린 공간을 줄이고 검은 삼목 판지단의 로지(일본 가옥에서 문안이나 뜰 안에 있는 좁은 통로)담을 희미하게 디자인하여 일본의 정통성을 표현하였다. 2층의 주거 공간은 노출콘크리트로 대비감을 표현하였다. 4.5m의 캔틸레버로 구조의 발코니는 마치 배와 같은 형상이다.

Architect : APOLLO Architects & Associates(Satoshi Kurosaki) **Location :** Shinagawaku, Tokyo, Japan **Site area :** 82.07m² **Building area :** 64.40m² **Gross floor area :** 123.02m² **Building coverage :** 78.47% **Floor space index :** 149.90% **Photographer :** Masao Nishikawa

설계 : 아폴로 아키텍츠 & 어소시에이츠(사토시 쿠로사키) **위치 :** 일본 도쿄 시나가와쿠 **대지면적 :** 82.07㎡ **건축면적 :** 64.40㎡ **연면적 :** 123.02㎡ **건폐율 :** 78.47% **용적률 :** 149.90% **사진 :** 마사오 니시카와

← View from Southeast ↰ Interior view of 2nd floor ↱ Seating area

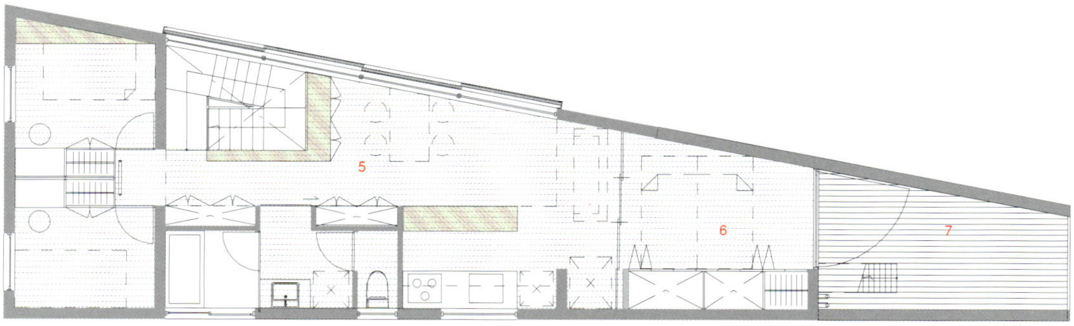

2ND FLOOR PLAN

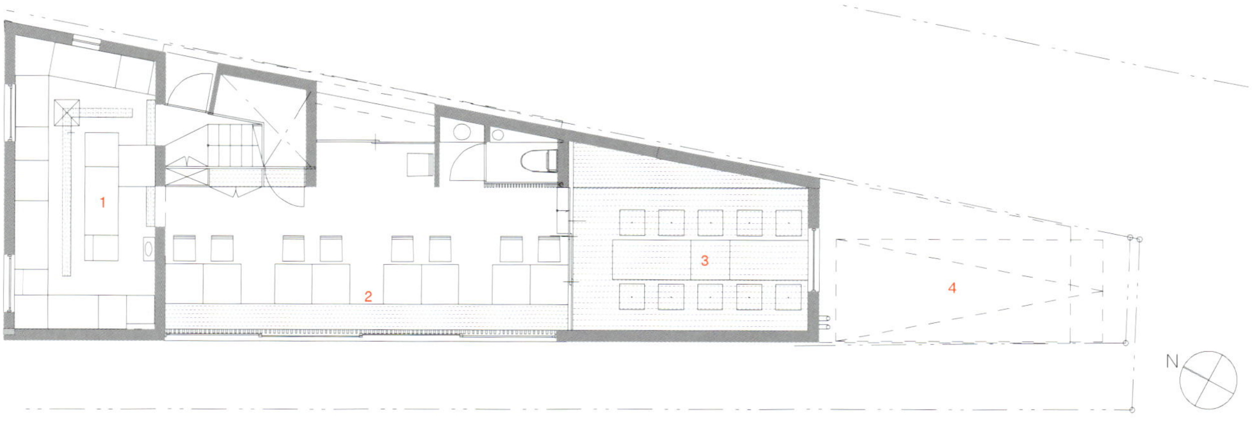

SITE / 1ST FLOOR PLAN

1 KITCHEN
2 BENCH SEAT
3 SEATING AREA
4 PARKING LOT
5 LDK
6 BEDROOM
7 BALCONY
8 ROOF BALCONY

← General view
↱ View of the 2nd floor
→ View from kitchen to balcony
↳ Bench seat area

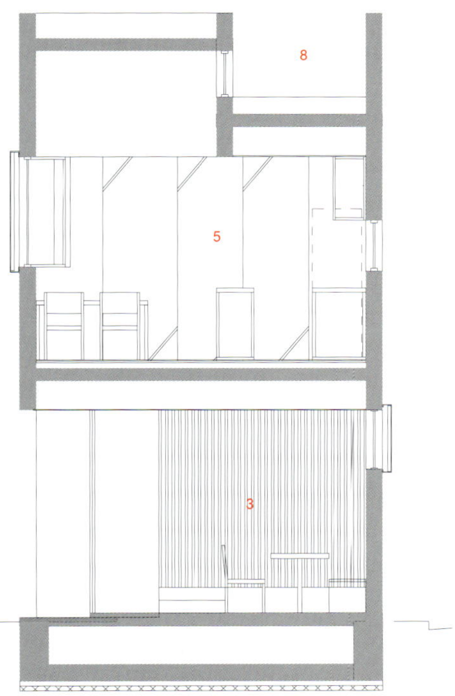

SECTION

BUSAN JUNG-GU YOUTH CULTURAL CENTER

부산 중구 청소년 문화의 집

The building is located a former printing alley. It was originally used as a community center and now is a culture space for youth. It has been there for about 25 years and displayed its history with the lapse of time. The youth culture center that is located on a small block stretched from a wide road stands on a dark asphalt street surrounded by achromatic-colored buildings. The project started from how to provide energy and events to this place and how to build relation between a building and people. We broke the existing dry urban spatial structure and rearranged factors to make a vigorous and active space. We focused on the visual display, flexibility and harmony of the building with surroundings rather than its usual features, hardness and independent existence. As the remodeling work was conducted, former repair works were revealed like tree rings and we also created new tree rings for the building in accordance with its new environments.

부산 중구 청소년 문화의 집은 예전의 인쇄 골목인 중구 보수동 2가에 위치한다. 원래는 동사무소로 사용되다가 청소년 문화공간으로 용도를 변경하였다. 25년이 지난 시간의 흔적이 묻어나는 장소이다. 큰 도로변에서 한 블록 들어와서 위치한 건물은 주변의 무채색 건물과 어두운 아스팔트 가로에 둘러싸여 있다. 이러한 장소에 활력을 불어넣기 위해 건축과 인간이 관계 맺는 것에서부터 계획을 시작했다. 이야기가 없는 건조한 기존 도시 공간의 구조를 깨트리고 재정리하여 활력과 생명력이 넘치는 장소를 만들고자 하였다. 건물의 일상적인 모습보다는 시각적인 강조로, 단단함보다는 유연함으로, 주변에 자리 매김보다는 드러날 수 있도록 했다. 리모델링을 위해 건물을 뜯을 때마다 알지 못했던 예전의 개보수 과정들이 마치 나이테처럼 드러났고, 그때마다 새로운 환경에 맞는 새 나이테들을 또 하나씩 만들어 나갔다.

Architect : Sangji Environment & Architects, Inc(Ko Seong Ryong) **Location :** Nampo-dong, Jung-gu, Busan, Korea **Site area :** 180.50m² **Building area :** 144.34m² **Gross floor area :** 494.49m² **Building coverage :** 79.97% **Floor space index :** 238.20% **Exterior finishing :** Aluminium board, High density wood panel **Design period :** 2008. 8 ~ 2008. 10 **Photographer :** Lee In Mi

설계 : (주)상지E&A건축사사무소(고성룡) **대지위치 :** 부산광역시 중구 보수동2가 64-1 **대지면적 :** 180.50㎡ **건축면적 :** 144.34㎡ **연면적 :** 494.49㎡ **건폐율 :** 79.97% **용적률 :** 238.20% **외부마감 :** 알루미늄 판, 고밀도 목재패널 **설계기간 :** 2008. 8 ~ 2008. 10 **사진 :** 이인미

← Exterior view ↑ View of the 2nd floor

↑ Roof garden ↓ Detail of facade → View from left side

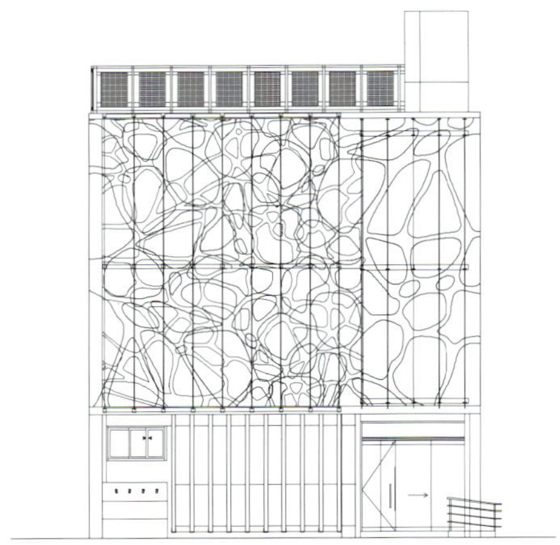

FRONT ELEVATION

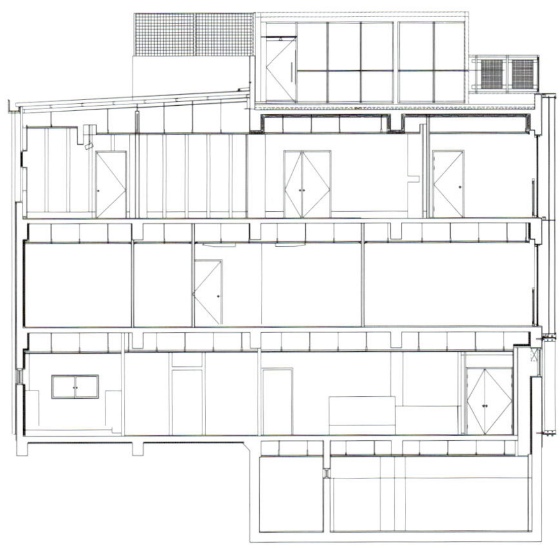

LONGITUDINAL SECTION

1 CHILDREN'S LIBRARY
2 LIBRARY
3 OFFICE
4 AV ROOM
5 HALL
6 CREATIVE SPACE

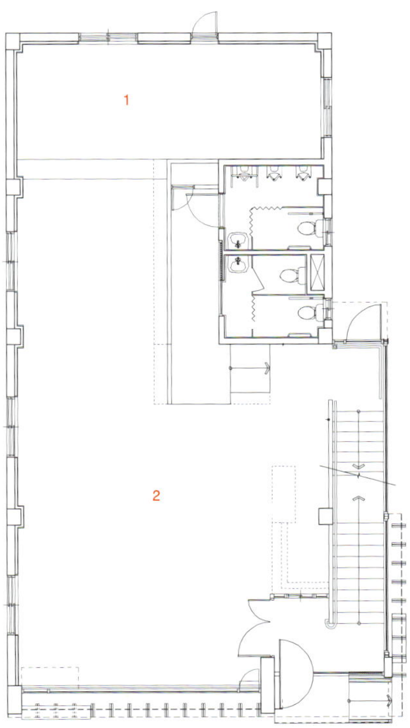

1ST FLOOR PLAN

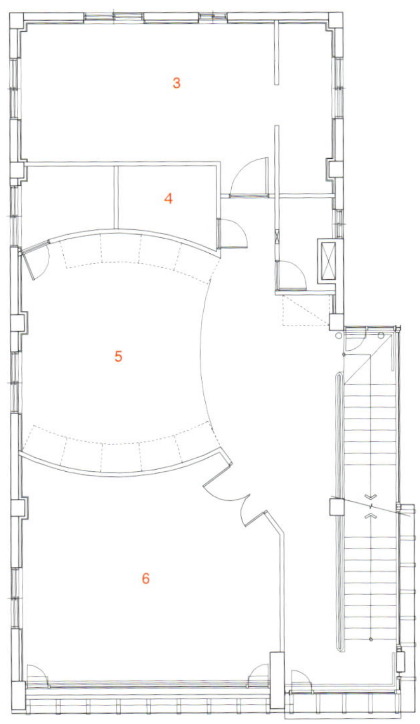

2ND FLOOR PLAN

← Stairs ↰ Stairs ↱ Exterior view

Y&M HOUSE

Y&M 하우스

Y&M House is located in the middle of the way from the main entrance of Seoul thermoelectric power plant to the front of the facilities, responding to the gentle urban context of the surroundings. Therefore, we organized the void without protruding or recessing, lowering or raising the volume, to be harmonized in the urban context and to be embraced by the order of the adjacent roads and surrounding buildings. The perfectly partitioned entire facade of the building and gneiss areas give a stable feeling to the form of the building. In addition, by mixing the strong texture of gneiss and the clarity of glass, we exposed the original texture of the materials to create a dramatic contrast. The vertically raised partition of the face expresses a controlled aesthetic with the Mondrian motif, and the oblique textured gneiss implies determinacy. As a buffering space between the external spaces and the internal spaces, the emotional space was planned. It is a metal fabricated space which responds to the urban order while giving a light but emotionally heavy spatiality to the space through the elevation.

Y&M 하우스는 서울화력발전소 정문에서 정면으로 이어지는 길 중간에 위치하며 주변의 점잖은 도시적 문맥을 이어간다. 도시 문맥과 어울리고, 도로와 주변 건축물의 질서에 포용되도록 볼륨을 돌출시키거나 후퇴시키지도, 낮추거나 높이지도 않고 그 빈틈을 구성해 나갔다. 건물 전면 전체와 편마암의 면 비례를 기하학적 황금분할로 구성하여 이상적이고 안정적인 형태감을 부여했다. 또한 편마암이 갖는 강한 텍스처와 맑은 유리를 같은 면에서 구성하여 극적인 대비감을 주는 등 재료가 가지는 질감을 그대로 노출시켰다. 수직 상승적인 면 분할은 몬드리안을 모티브로 절제된 미학을 표현하며, 사선 방향성을 갖는 강한 질감의 편마암 매스는 확정성을 내재한다. 외부 공간과 내부 공간의 완충적 공간으로 심리적 영역을 계획했다. 메탈 패브릭으로 이루어진 공간으로 면 요소를 통해 도시 질서를 따르면서 시각적으로는 가볍고, 심리적으로는 깊은 영역감을 부여했다.

Architect : CY Architecture(Youn Gyu Seob) **Location :** 362-16, 20 Hapjeong-dong, Mapo-gu, Seoul, Korea **Site area :** 286.50m² **Building area :** 142.74m² **Gross floor area :** 982.91m² **Photos offer :** CY Architecture

설계 : (주)씨와이건축사사무소(윤규섭) **위치 :** 서울특별시 마포구 합정동 362-16, 20 **대지면적 :** 286.50㎡ **건축면적 :** 142.74㎡ **연면적 :** 982.91㎡ **사진 :** (주)씨와이건축사사무소 제공

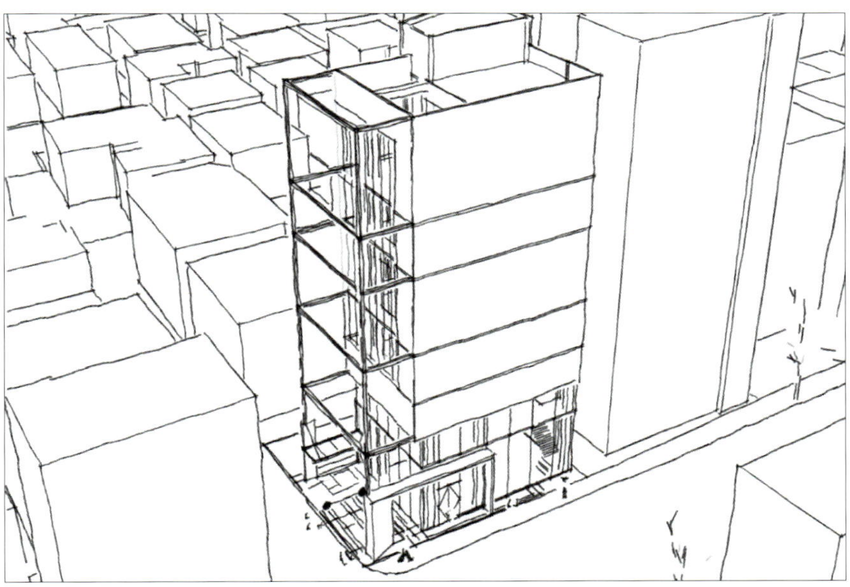

← Exterior view

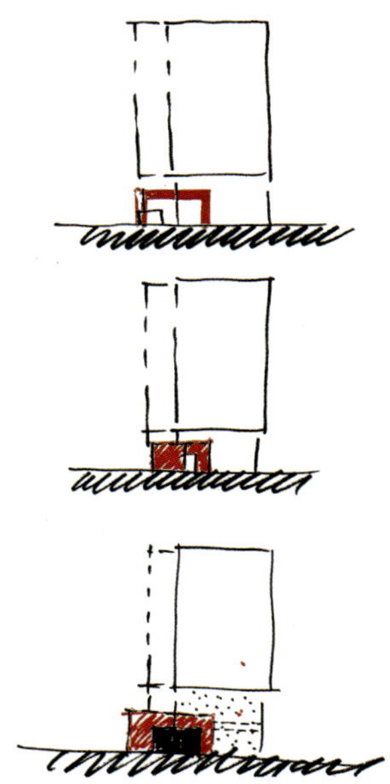

SKETCH

Sunken / View from the external stairs

Metal fabric area / Exterior view from below

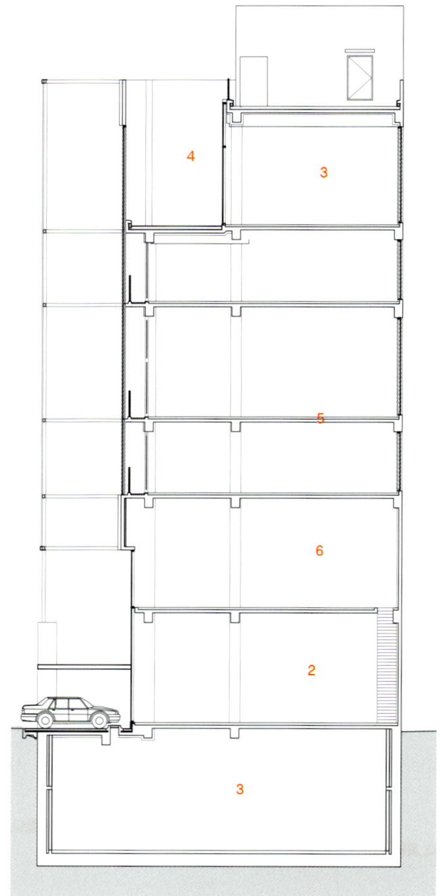

SECTION

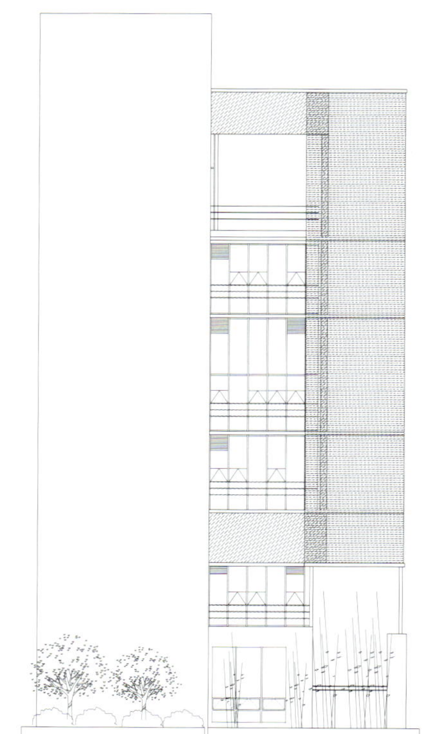

WEST ELEVATION

↰ View of the shop ↲ View of the shop ↑ View of the shop

1 HALL
2 SHOP
3 STUDIO
4 DECK

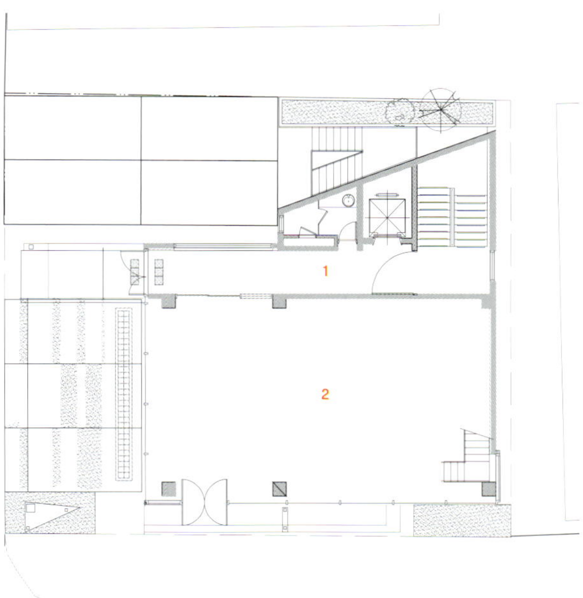

1ST FLOOR PLAN

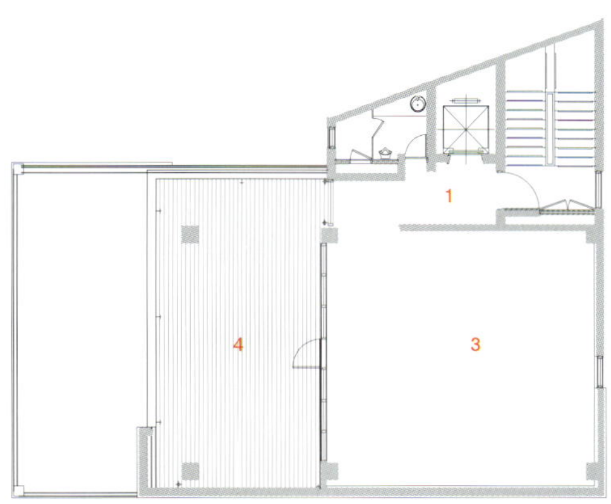

6TH FLOOR PLAN

THE YARDMASTER'S BUILDING

멜버른 조차장

Typically, a service building inaccessible to the public like this would be treated in a pragmatic and unremarkable way. In contrast, the Yardmaster's Building confidently presents itself as a 'public' entity. So that the building's public interface would not be contingent upon the inevitable machinations of this complex stakeholders mix, the planning was contained within a simple flexible shell. Internally, the building is robust and matter of fact. A rational system of glazing integrated with the repeated external pattern allowed each room to have at least one beautiful and exotic window. The external walls are made from a double-skin precast concrete system requiring a combination of production processes not attempted before. On the external face, a simple tiling pattern was formed and the relief polished. As it ages, the patina of the facade will express its environment, as if it had always been there or simply emerged from the ground. Windows are sealed from airborne particles and the fumes of the rail environment. This building is moody.

Architect : McBride Charles Ryan(Robert McBride, Debbie-Lyn Ryan) **Location :** Melbourne, Australia **Size :** 1,350m² **Photographer :** John Gollings

설계 : 맥브라이드 찰스 라이언(로버트 맥브라이드, 데비-린 라이언) **위치 :** 오스트레일리아 멜버른 **규모 :** 1,350㎡ **사진 :** 존 고링스

보통 일반인이 이용하기 어려운 공공건물은 실용성에 염두하고, 눈에 띄지 않는 형식으로 취급된다. 하지만 이 건물은 '공공' 건물의 위상을 당당하게 뽐낸다. 건물의 공공성에 관한 접점은 복잡한 이해 관계자들의 의도와 관계없이 단순하고 유연한 범위에서 계획했다. 건물의 내부는 튼튼하면서 실용적이다. 반복적인 외부 패턴으로 구성된 합리적인 창호 시스템은 각 내부 공간마다 아름답고 이국적인 창문을 가지게 한다. 외벽은 이전에는 시도되지 않았던 제작 과정이 요구된 이중 외피의 프리캐스트 콘크리트 시스템으로 만들어졌으며 광택이 나는 간결한 타일링 패턴으로 구성되었다. 시간이 흐를수록, 파사드의 그윽한 멋은 마치 건물이 그곳에 오랫동안 있었던 것처럼, 혹은 땅 속에서 그냥 솟아나온 것처럼 주변 환경에 어우러질 것이다. 창호는 공기중의 미립자와 기찻길의 매연을 막아준다. 조차장 건물에는 나름의 운치가 있다.

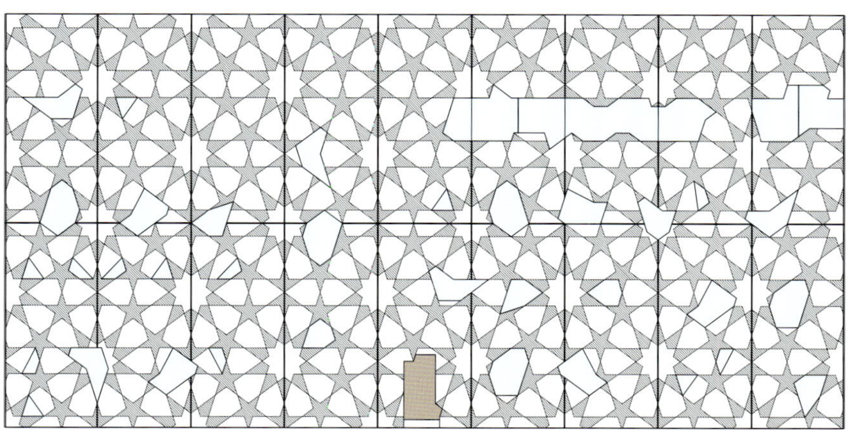

EAST ELEVATION

NORTH ELEVATION

← View from North

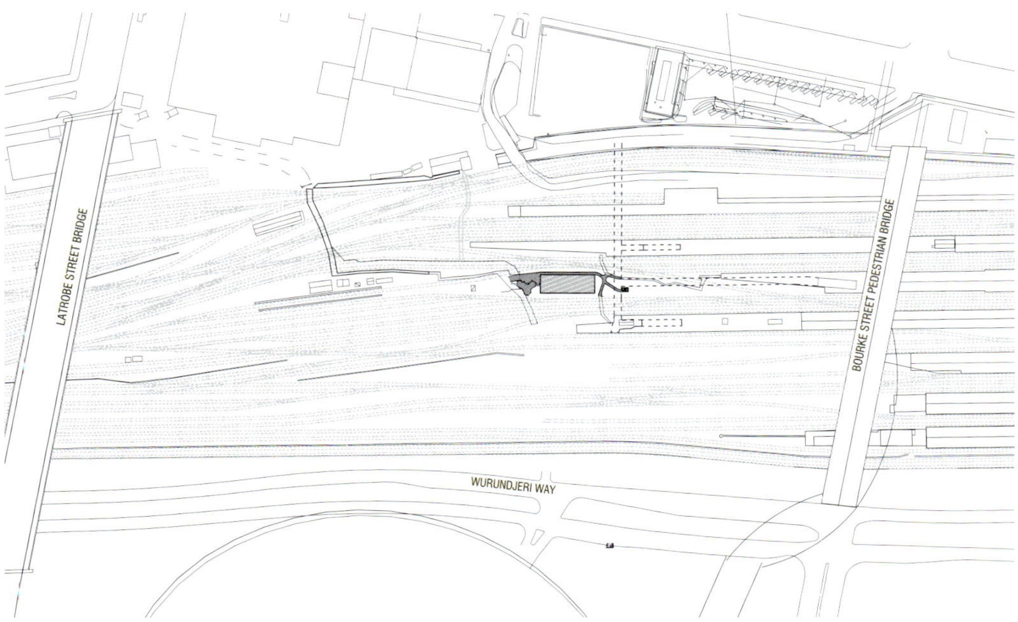

SITE PLAN

← Stairs
↱ General view
↳ Detail of facade

↑ View of the 3rd floor

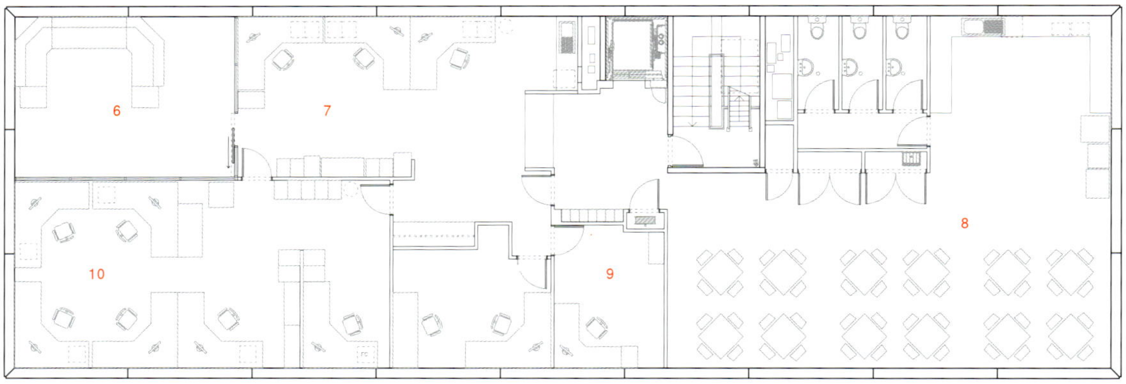

3RD FLOOR PLAN

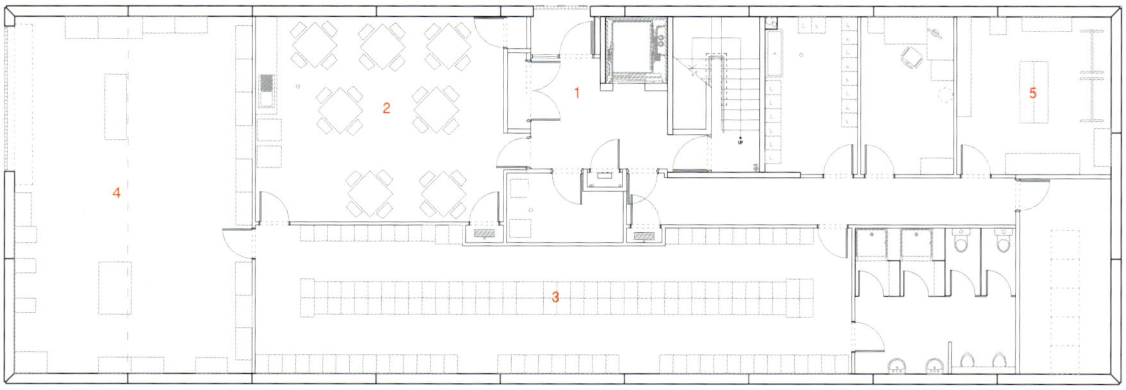

GROUND FLOOR PLAN

1 ENTRY
2 STAND-BY ROOM
3 LOCKERS
4 MAINTENANCE WORKSHOP / STOREROOM
5 DRIVERS STORE
6 SIGNALMEN OFFICE
7 YARDMASTERS OFFICE
8 RESTAURANT
9 SENIOR YARDMASTERS
10 CARS OFFICE

interior view
interior view
interior view

YAMAHA
MONT BLANC
PATEK PHILIPPE
GENEVE
Nisshindo
H&M
トップス
¥1,790
H&M
銀座フェミークリニック
www.femmy-c.com
メディカルエステ 銀座リセラ
9·10F
YAMAHA
YAMAHA
ROLEX

YAMAHA GINZA

야마하 긴자

To express the essence of Yamaha's brand, Japan's leading general musical instrument manufacturer, and materialize its architectural design, the basic themes of this architecture are set out to be "a building that exudes the feel of sound and music" and "the union of tradition and innovation". To achieve the building's concept image, which is a woodwind instrument that is placed inside a glass showcase, the connected three atrium spaces facing the streets are finished by wooden materials with curved moldings that resemble woodwind instruments. The facade of the building is composed of the diagonal grid expressing the fluidity of music, and the gold-leaf laminated glasses scattered in a random pattern expressing the brass instruments. With this facade, a number of musical senses are tried to be expressed in this architecture: a sparkling and shifting sense of sound, a rhythm, and a sense of constant change through time. Gold-leaf laminated glass, which was made by sandwiching Japanese traditional craftsman's gold dust between high-translucency glasses using modern technologies, represents Japanese images, in addition to an image of brass instruments. By hanging these gold-leaf laminated glasses with the diagonal cable-grid system, a new type of curtain wall like stained glasses was developed.

Architect : NIKKEN SEKKEI(Hozuma Chinio, Hiroyuki Shirai, Takeyoshi Suzuki) **Location :** Ginza, Tokyo, Japan **Site area :** 635.43m² **Building area :** 577.58m² **Gross floor area :** 7,582.99m² **Construction period :** 2007. 5 ~ 2010. 2 **Photographer :** Kenichi Suzuki Photography Studio, Osamu Tanaka

설계 : 닛켄 세케이(호조마 치노, 히로유키 시라이, 타케요시 스즈키) **위치** : 일본 도쿄 긴자 **대지면적** : 635.43㎡ **건축면적** : 577.58㎡ **연면적** : 7,582.99㎡ **공사기간** : 2007. 5 ~ 2010. 2 **사진** : 켄이치 스즈키 포토그라피 스튜디오, 오사무 타나카

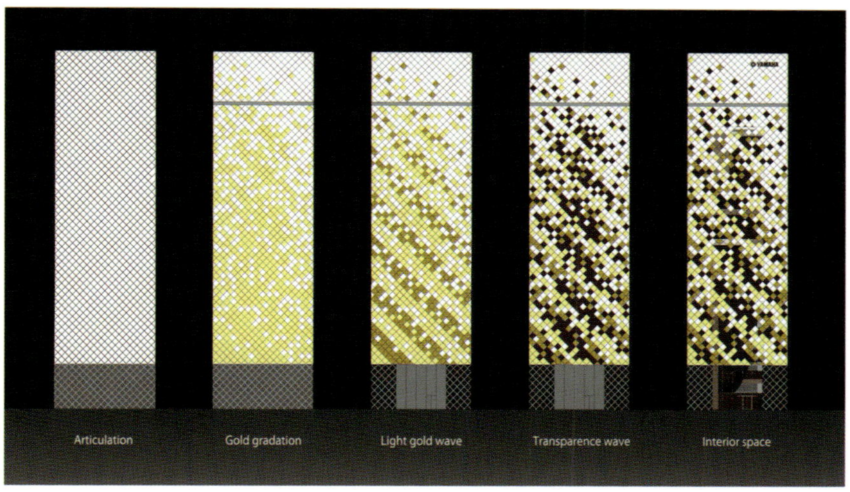

FACADE LAYER DIAGRAM

← View of the main facade

YAMAHA
MONT BLANC
PATEK PHILIPPE
GENEVE
FANCL
銀座七丁目
MONT BLANC

← Night exterior view ↰ Facade detail ↲ Music lesson room lounge ↑ Detail of facade from foyer

일본 악기 제조업체인 야마하 브랜드의 특성을 건축적으로 표현하기 위해 "소리와 음악의 느낌을 표현하는 건물", "전통과 혁신의 만남"을 주제로 접근했다. 거리쪽에 면한 세 개의 연결된 아트리움 공간은 유리장에 진열된 목관악기를 나타내는 곡선의 목재로 마감했다. 메인 파사드는 음악의 유연함을 표현하는 사선 격자무늬와 금관 악기를 표현하는 무작위로 흩어진 접합유리로 구성했다. 이로써 소리의 빠른 움직임과 감각, 리듬 그리고 시간에 따라 변화하는 느낌 등 음악적 감각이 건물 외관에서부터 표현된다. 금박을 여러 장 붙여 만든 유리는 금관 악기를 표현함과 동시에 일본적인 느낌이 묻어난다. 이와 같은 유리를 사선 케이블-그리드 시스템에 걸어 스테인드 글라스와 같은 새로운 형태의 커튼월이 만들어졌다.

← Entrance facade
↱ Music lesson room lounge
↳ Grand stairs of music school

1 GRAND STAIRS
2 MUSIC SCHOOL
3 YAMAHA HALL
4 FOYER
5 SHOP
6 MUSIC STUDIO

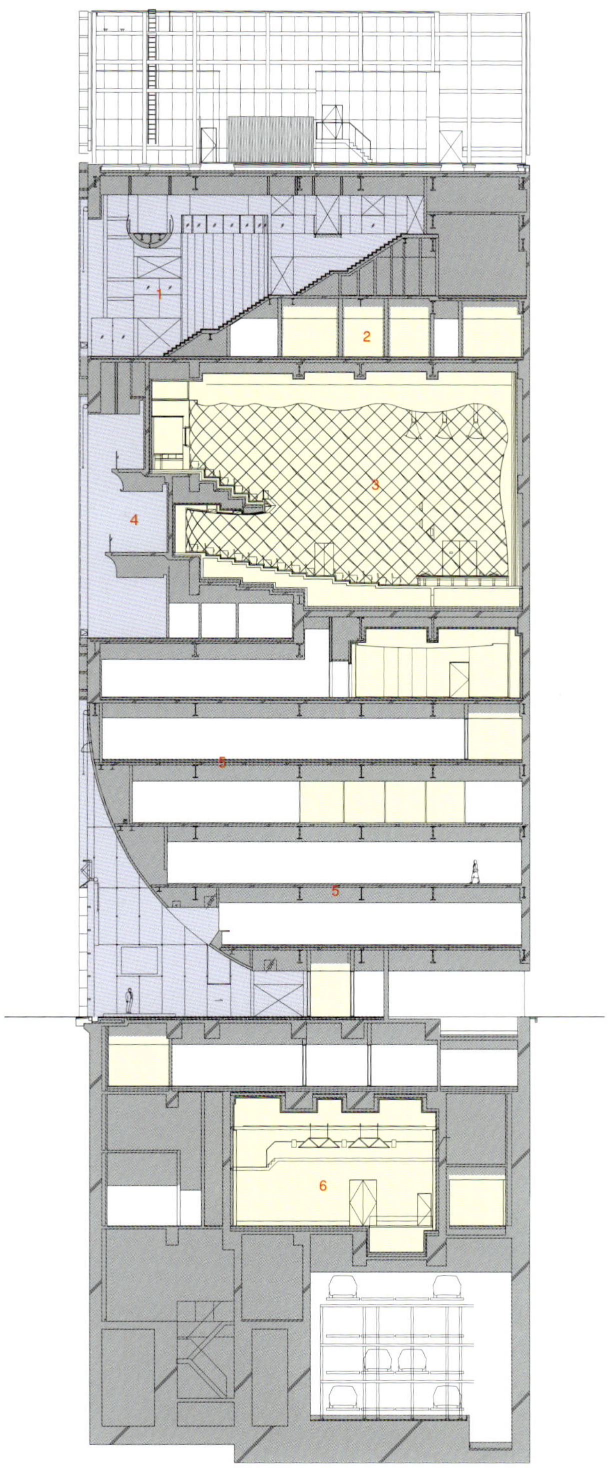

SECTION

↑ Yamaha hall ↱ Piano shop ↳ Salon

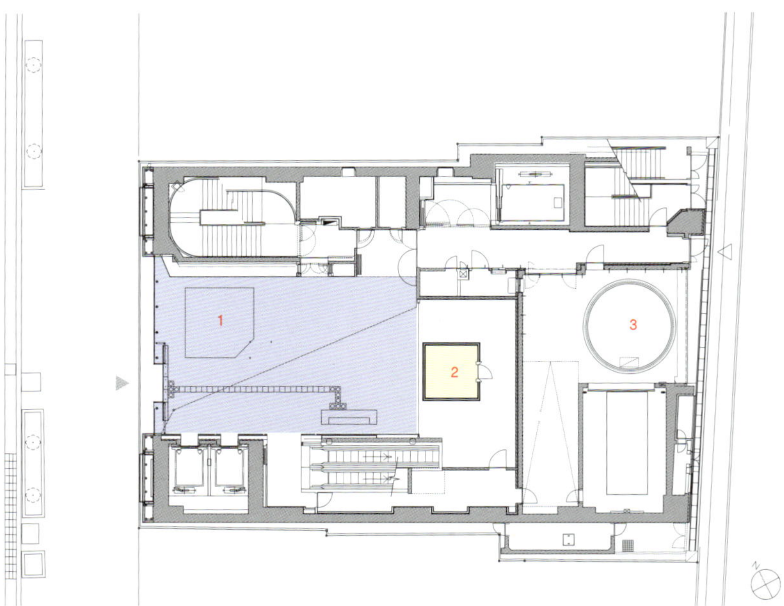

GROUND FLOOR PLAN

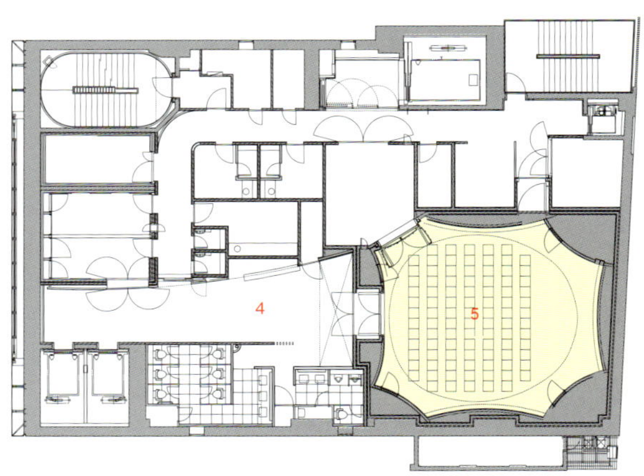

6TH FLOOR PLAN

1 PORTAL
2 ELECTONE ROOM
3 PARKING
4 LOBBY
5 SALON

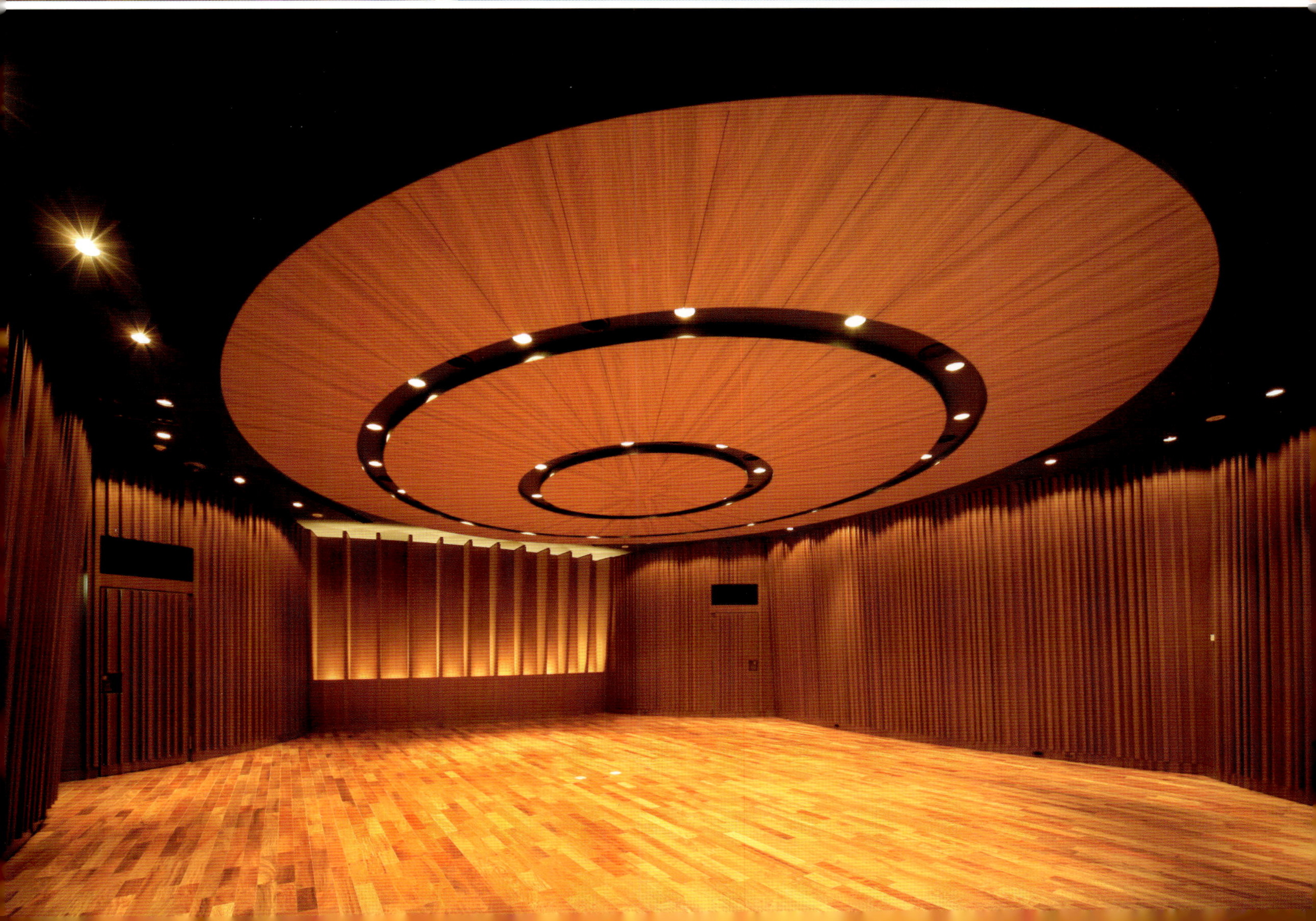

CHARLOTTE BUILDING

샬럿 빌딩

The Charlotte Building is a new-build speculative office project on the site of a former 1950s office scheme. Located in the heart of London's 'Noho' district, the Charlotte Building sits just behind the junction of Oxford Street and Tottenham Court Road on a tight urban site organized around the historic street pattern and roughly triangular in shape. Closely surrounded by buildings of differing ages and adjacent to the Hanway Street Conservation Area of semi-industrial Victorian buildings, the form of the building was required to respond directly to the site configuration - its triangular core located against the south-east boundary, creating a new landscaped courtyard and pedestrian route. Comprising approximately 4,545m² Net Internal Area and situated over 7 floors, the office floor plates are designed to maximise the flexibility of use for the office space and are based on a 1.5m planning grid to allow for cellular and open plan configurations.

런던 노호 지구의 중심부의 샬럿 빌딩은 1950년대식 오피스 건물이 있던 대지에 지어졌다. 역사적인 거리 패턴을 중심으로 옥스포드 거리와 토트넘 코트 로드의 교차 지점의 삼각형 대지에 위치한다. 각각 다른 시대에 지어진 건물들에 둘러싸여 있고, 반산업 시설인 빅토리아 시대의 건물들이 들어찬 한웨이(Hanway) 거리에 인접해 있어 대지 형태에 직접적으로 대응해야 했다. 남동쪽 경계선을 등지고 위치한 삼각형 핵심부는 색다른 경관의 뜰과 보도를 제공한다. 실내 면적 약 4,545㎡의 7층으로 구성된 오피스 층은 공간 이용의 유연성을 최대화시키고, 내부 공간은 열린 구조로 1.5m 그리드를 기반으로 한다.

Architect : Lifschutz Davidson Sandilands(Jackson Coles) **Location :** London, UK **Completion :** 2009. 12 **Photographer :** Chris Gascoigne

설계 : 라이프슈츠 데이비슨 샌디랜즈(잭슨 콜스) **위치 :** 영국 런던 **완공 :** 2009. 12 **사진 :** 크리스 가스코인

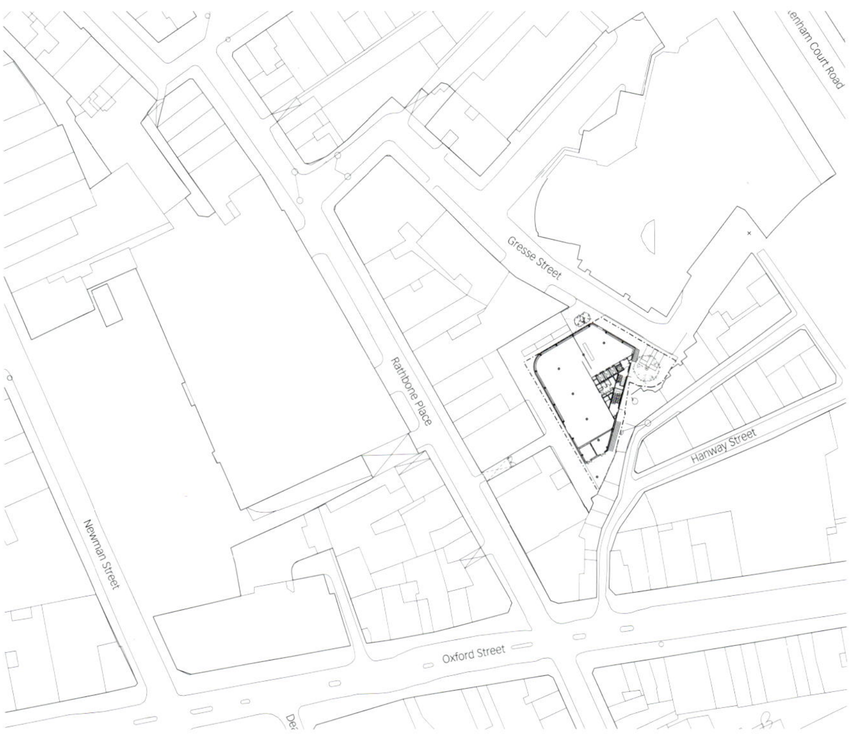

SITE PLAN

← Exterior view

← Street view ↑ Self-contained / alternative entrance → View from Rathbone Place

NORTH ELEVATION

EAST ELEVATION

↑ Reception area ↲ Reception area └, Bathroom → Skylight on the 6th floor

1 PARKING LOT
2 RECEPTION
3 OFFICE FLOOR
4 DISABLED PARKING LOT

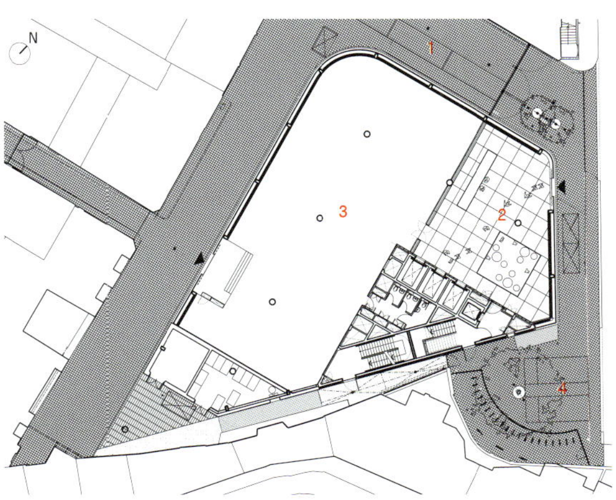

GROUND FLOOR PLAN

6TH FLOOR PLAN

↑ Typical office ← Terrace of the 6th floor

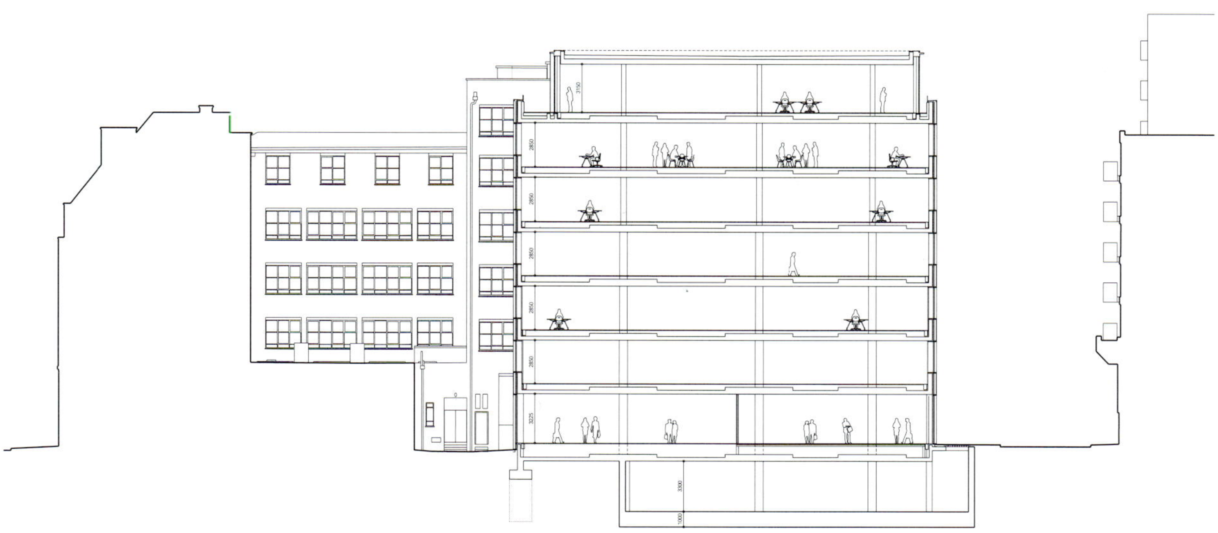

EAST-WEST SECTION

Meister Motors

DAECHI-DONG NEIGHBORHOOD FACILITY

대치동 근린생활시설

Around the site are low-rise buildings. The program idea was more important than the potential of the site on that land. When the given space program was organized, spaces were created and necessary process was taken, the key for the architecture was surface. How to deal with surfaces, what forms surfaces and what relation surfaces have with given programs becomes main issues in the beginning of the design. On the 1st and 2nd floors, there are spaces for car exhibition and sales, while from the 3rd floor to the 6th floor are rental offices. They intended to create one unified image by combining two different characterful programs. One powerful and continuous surface was folded to make the lower part a car exhibition hall with transparent void and the upper part rental offices with solid volume. The design idea was that the pattern of the front upper part's skin goes to ceiling and rear of the car exhibition hall and the exterior. Also, thick and folded side was expanded to suggest clear architectural gestures and provide a basic frame for lighting.

대지 주변은 평범한 저층 상가 건물들이 늘어서 있다. 대지의 강한 잠재력 보다 프로그램의 아이디어가 우선되는 땅이라 생각했다. 주어진 스페이스 프로그램을 조직하고, 공간을 만들고, 제스처를 발생시키는 일련의 과정 중에서 건축적 수단으로 Surface(연속 면)을 이용했다. 면을 다루는 방식과 그를 구성하는 재질, 주어진 프로그램과의 관계 등이 설계 초기의 주된 이슈가 되었다. 건물은 1, 2층에 자동차 전시 및 판매, 3~6층의 임대 사무실로 구성된다. 두 개의 다른 성격의 프로그램을 잘 조합하여 하나의 통합된 이미지를 부여하고자 했다. 하나의 강력한 연속 면을 접어서 이를 매개로 저층부는 외부처럼 느껴지는 투명한 보이드의 자동차 전시장으로, 상층부는 임대 사무실로 솔리드한 볼륨으로 만들었다. 전면 상층부 외피의 패턴이 자동차 전시장의 천장과 후면, 바닥을 타고 외부까지 연장되는 아이디어를 생각하고 설계했다. 측면은 접힌 면을 깊이감을 가진 두께로 확장하여 건축적 제스처를 확실히 가지도록 하였고, 경관 조명의 기본 틀이 되도록 하였다.

Architect : urbanEx Architects & Associates(Oh Seom Hoon) **Location :** 949-3 Daechi-dong, Gangnam-gu, Seoul, Korea **Site area :** 1,280.00m² **Building area :** 632.35m² **Gross floor area :** 6,384.70m² **Building coverage :** 49.40% **Floor space index :** 247.87% **Construction period :** 2008. 2 ~ 2008. 12 **Photos offer** : urbanEx Architects & Associates

설계 : (주)건축사사무소 어반엑스(오섬훈) **위치 :** 서울특별시 강남구 대치동 943-3외 4필지 **대지면적 :** 1,280.00㎡ **건축면적 :** 632.35㎡ **연면적 :** 6,384.70㎡ **건폐율 :** 49.40% **용적률 :** 247.87% **공사기간 :** 2008. 2 ~ 2008.12 **사진 :** (주)건축사사무소 어반엑스 제공

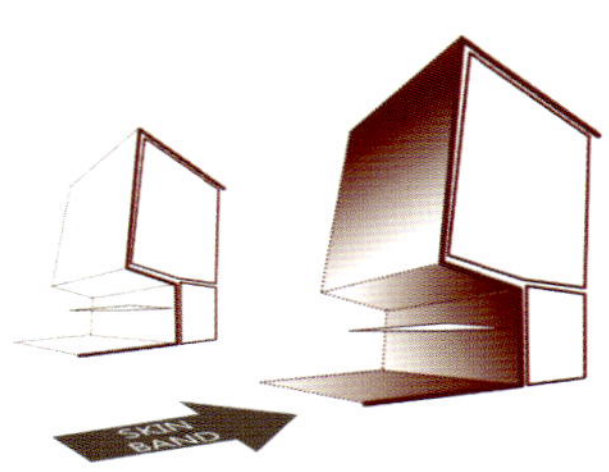

SKIN BAND DIAGRAM

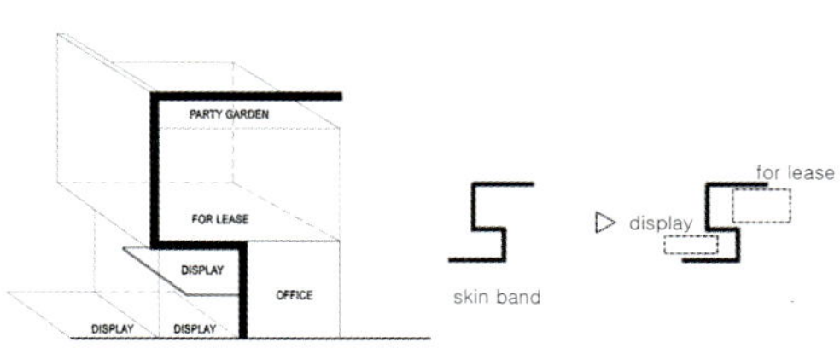

SPACE PROGRAM

← Exterior view

↑ Night view ↱ Showroom ↳ Void space

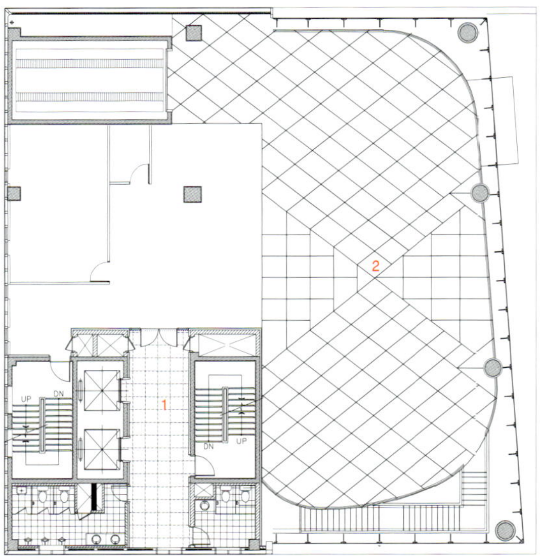

2ND FLOOR PLAN

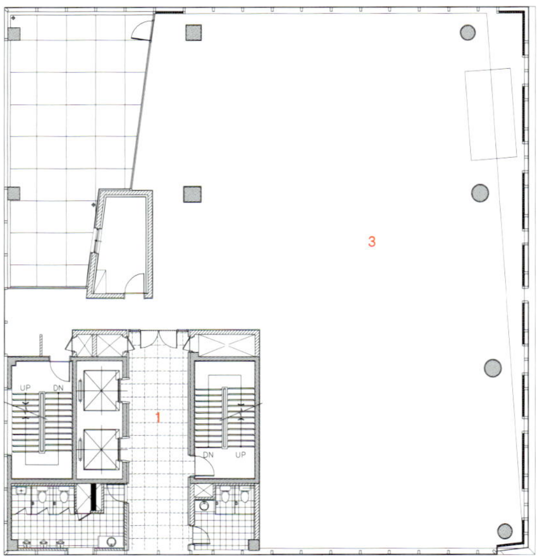

3RD FLOOR PLAN

1 HALL
2 SHOWROOM
3 OFFICE

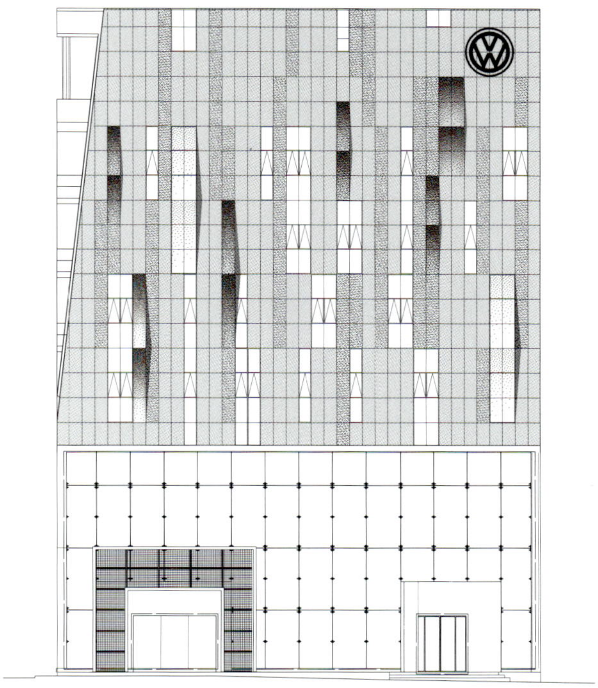

FRONT ELEVATION

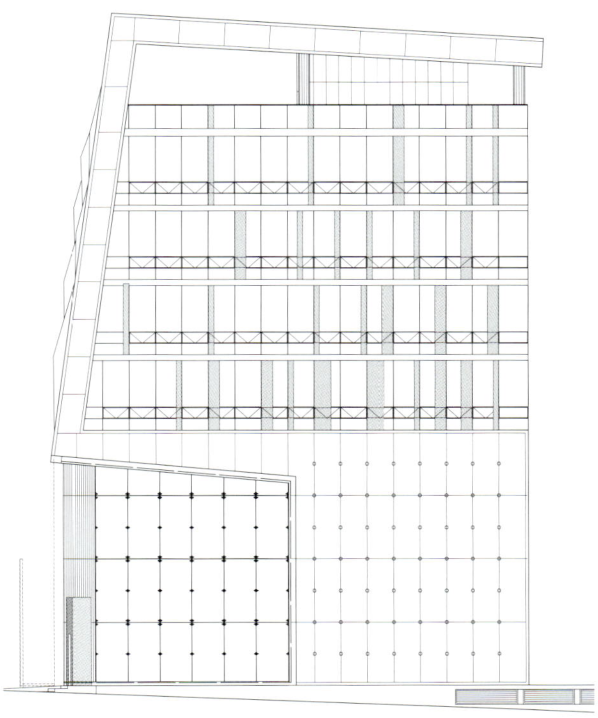

RIGHT ELEVATION

LA-CUBO RETAIL

라꾸보

The Cheonan Recreation Complex, where LA-CUBO RETAIL is located, is the only living facility within the Lake Park at the center of the complex. It has the optimal location as a commercial facility since the building faces Cheonan Culture & Art Hall while having the lake and the open-air stage as the background. On the other hand, it has to express its own identity in the surroundings of the large scale buildings. LA-CUBO is a compound which means a 'cube' or 'diced material' in Italian. It may be originated from the extraordinary form of the building. The building was designed to meet the requirements as a commercial building, utilize the surroundings, and create indoor/outdoor spaces to attract as many visitors as possible. The different indoor/outdoor spaces which are created by the two ㄷ-shaped plates are organically connected to other spaces for various views. Despite of its small scale, the building is expected to play a role of a 'cultural icon' in the future as it is placed at the center of the complex.

라꾸보는 천안종합휴양지 중앙에 위치한 호수공원 부지 내에 유일한 근린생활시설이다. 호수와 야외 무대를 배경으로 워터파크, 천안종합문화예술회관과 마주보고 있는 상업시설로 단지 내 최적의 입지 조건을 갖추고 있다. 하지만 주변의 대형 시설물 틈바구니에서 자신만의 몸짓을 표현해야 했다. 라꾸보는 이탈리아 합성어로 입방체의 뜻을 가지고 있으며 "어떤 재료를 네모썰기 한 것"이라는 의미도 지니고 있다. 이는 건물의 기이한 형태에서 비롯되었다. 본 건물은 상업시설이 가지는 요건과 입지조건, 또 주변 환경을 최대한 이용하여 다양한 배경을 보여줄 수 있는 많은 조망과 시설 이용객들을 많이 집객할 수 있는 실내외 공간 창출을 중점으로 고려하여 계획하였다. 두 개의 "ㄷ"자 형태의 단순한 선형의 판들이 서로 엇갈리면서 만들어내는 다양한 실내외 공간들은 외부와 내부를 서로 유기적으로 연결하면서 다양한 경관을 보여준다. 작은 덩치이지만 향후 단지 내 중심적 위치로서 색다른 "문화 아이콘"이 되기를 기대한다.

Architect : JUNGLIM Architecture(Kim Bok Soo, Kim Kyung Hoon, Jung In Hwan, Kim Hae Jin) **Location :** 671 Yongwon-ri, Seongnam-myeon, Cheonan, Korea **Site area :** 1,000.1m² **Building area :** 394.59m² **Gross floor area :** 998.50m² **Building coverage :** 39.46% **Floor space index :** 99.84% **Photographer :** Namgoong sun

설계 : (주)정림건축종합건축사사무소(김복수, 김경훈, 정인환, 김해진) **대지위치 :** 충청남도 천안시 동남구 성남면 용원리 671번지 **대지면적 :** 1,000.1㎡ **건축면적 :** 394.59㎡ **연면적 :** 998.50㎡ **건폐율 :** 39.46% **용적률 :** 99.84% **사진 :** 남궁선

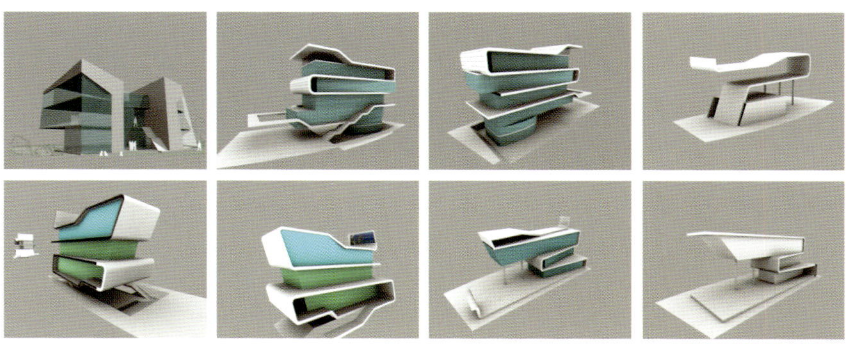

VOLUME PROCESS

← View from Southeast

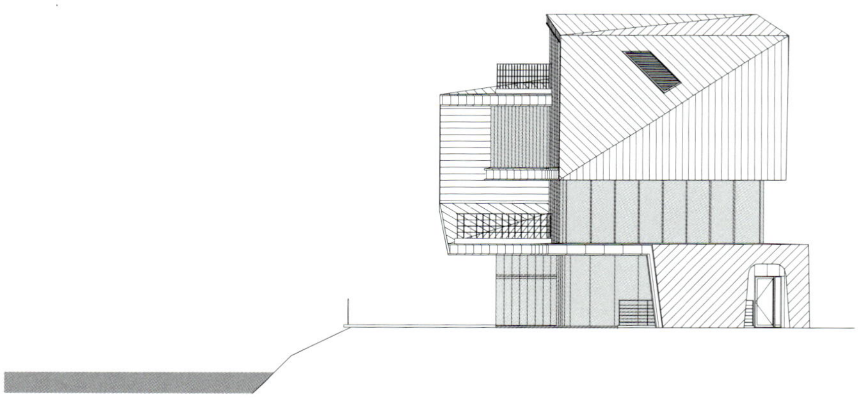

NORTH ELEVATION

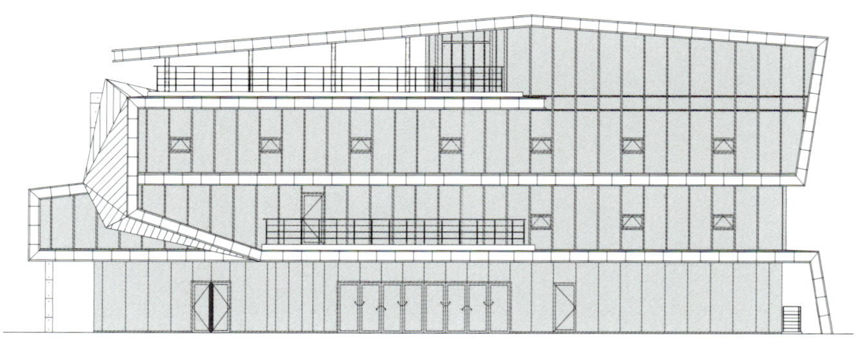

EAST ELEVATION

↑ View from Southeast
↑ View from Northwest
→ View from West

← Exterior view → Exterior view

↑ View from South ↲ Floating volume with deck space ↳ Rendering image → The part of exterior view from East

VOLUME IMAGE

la ubo
la ubo

← Interior space that two floor open ↑ Inetrior space

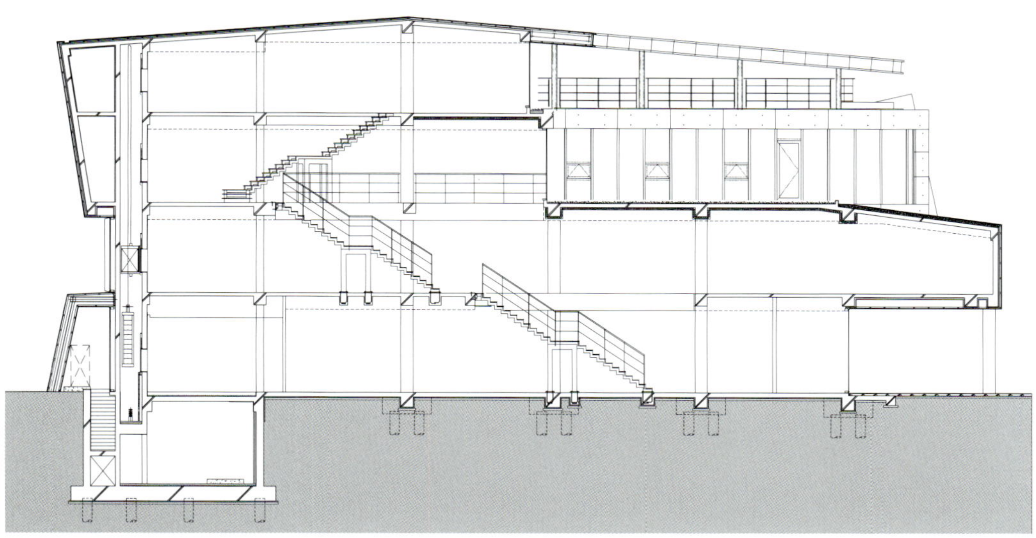

SECTION

↑ Roof deck
↵ Detail of facade
↳ Deck of the 2nd floor

↑ Restaurant of the 1st floor ↓ Stairs

1 RESTAURANT
2 DECK
3 HALL

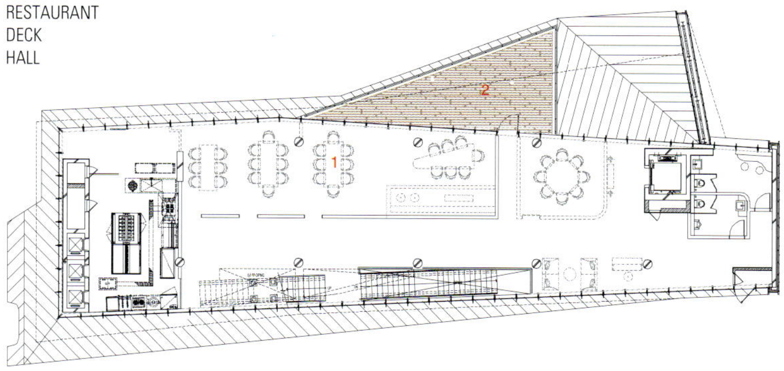

2ND FLOOR PLAN

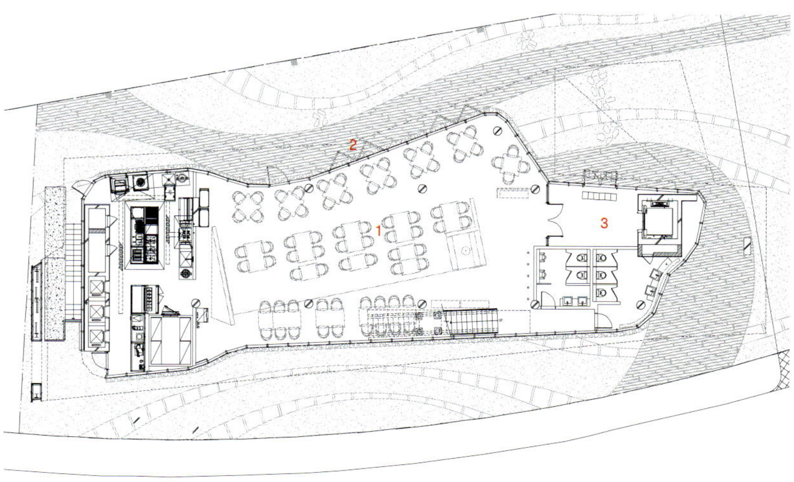

1ST FLOOR PLAN

CHASM

캐즘

The specific site for this combined programs of private architecture, relatively compact in its size, takes one of the regularly sliced lots in a quite suitably established suburban compound of similar kinds. Unique and exclusive in such aspects as its scenes and settings, objectives, programs, contents and even itsoccupants, this village named Heyri already has become an internationally acclaimed cultural community. The place itself apparently did not provide the clues enough to launch thoughts and ideas for the journey of rendering forms and images in this scheme. Neither any specific tensions or references around the site seemed to be perceivable righthand, nor any preliminary contexts or indications to interpret at the initial stages of the design. Put in other words, the given lot alone might as well even be regarded just as a blank margin among the 'blank' in terms of peripheral or external approaches to this project. Now as part of a subsequent and alternative measure, and as is mostly typical with this type of deadlock an internal way of logics and reasoning had to be brought up in my mind. Linguistically driven development of forms, images, and spatial languages as well as interpretations of the given program constantly intervene in the materializing procedure of the scheme; utilizing the fundamental 'thematic' term as the primary generator. The term I forged this time as the theme word was 'chasm' to integrate all the issues and requirements, and even limitations and restrictions latent within the program and the place. Entire process was supposed to be anchored upon its own group of meanings referring to such words as alienation, breach, cavity, cleavage, hollow, opening, hiatus, skip, gorge, void, split, etc. I suppose that three different discrepancies, implications or narratives intrinsic in this particular incident of architecture provided major sources to personally coin that word at the beginning. The community that this site is registered to initially maintained some ten years ago the notion that it has an extremely idyllic setting where these days there appears to be some missing senses of relativity, among the proposed natural topography, open spaces, verdure, visual distinctness, rural life, vegetations, sense of occupancy, building density, and even the communal activities and events within, etc. Presumably a quite potential amount of friction, incompatibility, interferences, alienations lies even among these proposed programs-residences, plats, workshops, exhibition spaces, offices, coffee and wine stands, etc regardless of their apparently manageable sizes and figures. Distinctively enough, the owner couple, so much sophisticated individuals, for this commission tend to exhibit highly clear senses and notions of distances, or furthermore of alienations in some areas from the capitalistic banality or vanity flooding on the current society. Variety of chasms derived from the symbolic, semantic and experiential terms here are supposed to be practically put under the process of transcriptions to actual visual and spatial languages and terms. Mainly peripheral devices to generate immediate senses of spaces could be recognized as a few mentionable output of the transcriptional process of the thematic term 'chasm' ; such as a large linear reflecting pool cutting across the entire structure, three different approaches indicated through irregular cleavages, relatively tiny, hidden slits in northern facets, two of the larger cutaway openings in southern facets, three different types of decks exclusively open upward to the sky, a corner protrusion of the window, two top-lights, an upper floor bridge between the two 'alienated' masses,

Architect : asylum(Kim Hun) **Location :** 1652-168 heyri art-valley, Paju, Korea **Site area :** 480.1m^2 **Building area :** 179.11m^2 **Gross floor area :** 469.06m^2 **Building coverage :** 37.31% **Floor space index :** 76.31% **Construction period :** 2006. 6 ~ 2007. 1 **Photographer :** Park Wan Soon

설계 : 어싸일럼(김헌) **위치 :** 경기도 파주시 탄현면 **대지면적 :** 480.1㎡ **건축면적 :** 179.11㎡ **연면적 :** 469.06㎡ **건폐율 :** 37.31% **용적률 :** 76.31% **공사기간 :** 2006. 6 ~ 2007. 1 **사진 :** 박완순

View from Southwest View from Northeast

↑ General view → Reflecting pool

의사-전원(擬似-田園)풍의(pseudo-idyllic) 어딘가 균질한 세팅, 이를 바탕으로 만들어지는 하나의 '의사(擬似)-마을'. 그 테두리 내에 분뎡 초점을 잃은 듯한 유사 프로그램들이 연이어 시각화 되어 가는, 그럼에도 이미 명소가 되어 버린 곳. 장소는 그 곳을 후경으로 갖는다. 전원이라 스스로 칭하는 바와는 어긋나게 이 '마을' 속에 처한 대지는 도시적 본능으로, 엄밀함과 정연함으로 재단 된 형상들 중 하나일 뿐이다. 주변 또한 그저 느슨한 관계나 질서의 맥(脈)(con-)들, 보편을 넘지 못하고 마는 해석의 실마리(絡)(-texts)들 따위만 그 희미한 자취를 읽게 한다. 외적으로 맞물려야 할 관계나 질서의 톱니바퀴가 뚜렷이 규정되기 어려운바, 어떤 의미로든 이 작업이 발생키 위한 장소란 애초 공백 속의 여백에 가까웠던 것이다. 이 사건 역시 구축이라는 필연에 나서기 위한 장소 본연의 레퍼런스 내지는 발상의 초기 단서들이 외부로부터 대거 가셔져 있음을 말함이다. 불가피하게 외부와의 절연, 이어 언어중심의 사적, 내적 질서들을 추출해보는 자생이란 대응 모드가 개입한다. 곧 하나의 테마 어휘 - 공간, 물질, 시간, 현상 등에 입혀지는 의미들의 '균열' 또는 '벌어짐'(chasm) - 가 이 작업의 구축을 위한 산모로 불러들여진다. 장소가 처한 분열적 양태나 프로그램에 선험적으로 뚜렷이 함유된 이질 내지는 이격의 성분 또 이 공간의 점유 주체로 설정된 두 인물, 그들과의 교류를 통해 시사된 그들 나름의 공간 인식이나 개인적 사회적 의식의 간극 등, 몇 가지 갈래를 따라 이 어휘는 효과적으로 계열화하는 것이다. 장소적 의미들의 표면에 새겨진 주된 균열이라면 이 기획마을의 풍경이 처한 당착이다. 성장할수록 스스로의 세포들에 의해 그 투사된 정경이 눈에 띄게 잠식당하는 있는 필연을 말함이다. 아마 풍경과 밀도와의 관계에 대해 초기에 지나치게 낙관했던 기획이었을 지도 모른다. 또 이 작업에 제안된 프로그램들 간에 본질적으로 발생하는 의미와 경험의 골짜기 역시 테마 어휘(chasm)의 울림, 그 반사파들을 발하고 있다. 비교적 소박한 몸집 내에 주거 행위와 식음, 개인의 지적 생산, 작업, 전시, 기타 미래의 문화적 행위 등이 엉겨질 그 불안한 동거는 끊임없는 '거리감'의 비트들로 변환되어야 하는 것이다. 이 작업을 의뢰한 인물들 역시 여러 면모로 현대사회가 좇는 잡다한 보편적 가치들로부터 간단히, 한편으로 고요히 물러서 있는 존재들이다. 그것이 능동적인 제스처든 냉담함이든 아니면 소외됨이든 그들 특유의 정서와 외부 일상의 분주함 사이엔 깊게 패인 골(chasm)이 자리하며, 둘 사이 하등 화해의 여지는 보이지 않는다. 장소와 프로그램에 발생한 상징적, 경험적 의미들의 균열은 틈새, 벌어짐, 거리, 함몰, 탈락 등의 시각 언어들로 파생하며 각기의 공간 언어들에 전사(轉寫)된다. 몸체를 가르는 반사 못, 벌어진 틈새로 지시되는 세 개의 진입 공간, 북측 면에 찢긴 네 개의 세장한 홈집, 남측 면에 비교적 크게 도려내어진 두 개의 개구부, 3층의 상부로만 열린 세 가지 타입의 외부 공간, 모서리에 돋아난 창, 두 개의 천창, 상부 브리지 등. 일련의 주름, 파열, 상처, 흠집 등으로 군살이 다져지고 부풀어 오른 상피는 그 속에 여러 형식의 임계적 장치들을 품게 된다. 이는 내 · 외부 공간의 일상적 경험 속에서 수시로 우리 의식의 수면에 미세한 파동을 일으키게 할 일종의 과도적 장치들이다. 공간의 내 · 외부가 접하는 임계 부근에 삽입된 그것들은 내부가 해소되었다고 믿어지는 영역의 바깥을 또 다른 스케일, 형식의 내부 공간으로 다시 잇는다. 외부 공간의 안쪽 접면은 쉽게 내부에 이르지 못하고 이와 다른 차원의 외부 공간이란 배타적인 필터를 거쳐야 한다. 구체의 외피에 가해진 균열과 상처들로 공간의 내 · 외부는 끝없는 반목과 화해와 갈등의 시간을 견디고 있는 것이다.

← Exterior view ↑ View from the bridge ↵ Exterior view ↳ Stone deck

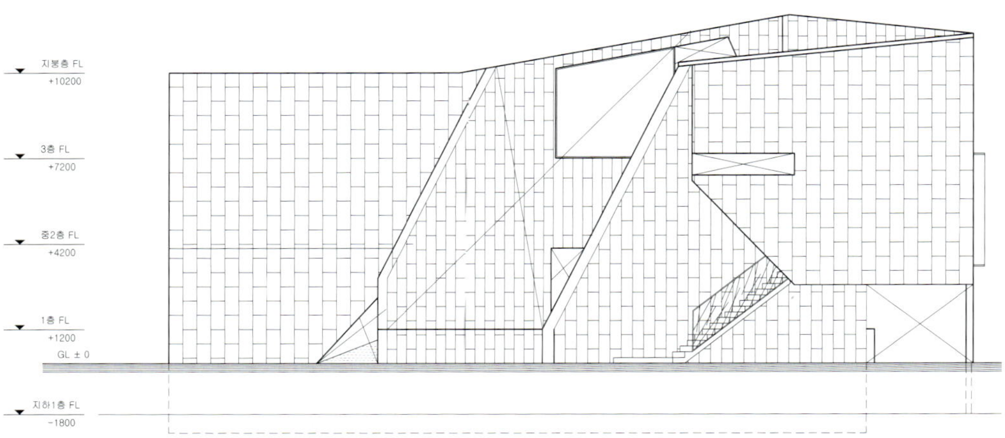

NORTH ELEVATION

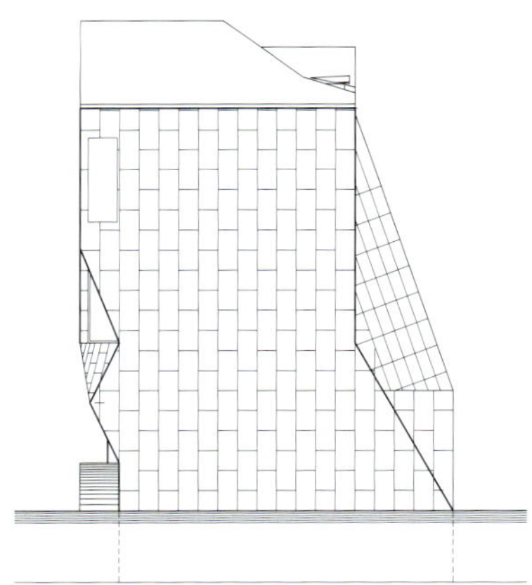

EAST ELEVATION

↰ Deck on the 3rd floor open upward to the sky ↑ View from the living room to the library ↓ Living room

↑ Interior view of the 1st floor

SECTION

MODEL

MODEL

⌐ Interior view of the 1st floor ↑ Gallery on the basement ↓ Gallery on the basement

1 TOP LIGHT
2 LIVING ROOM
3 STONE DECK
4 LIBRARY
5 OFFICE
6 OPEN ABOVE
7 POOL
8 PARKING LOT

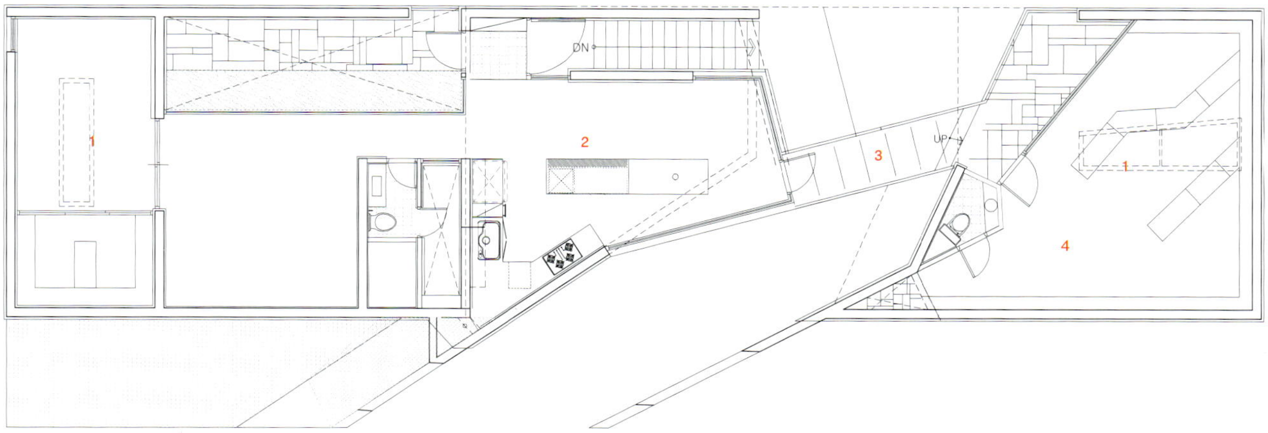

3RD FLOOR PLAN

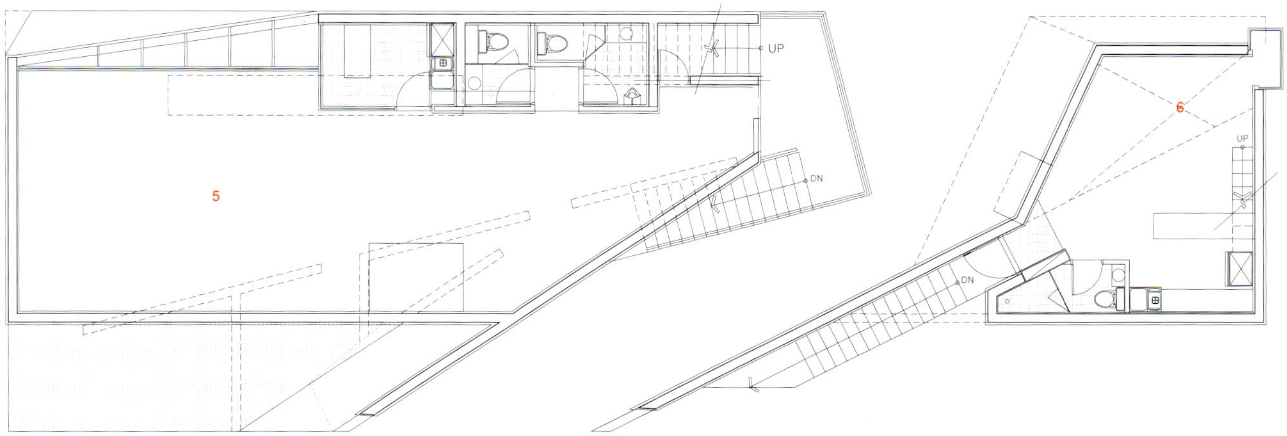

2ND FLOOR PLAN

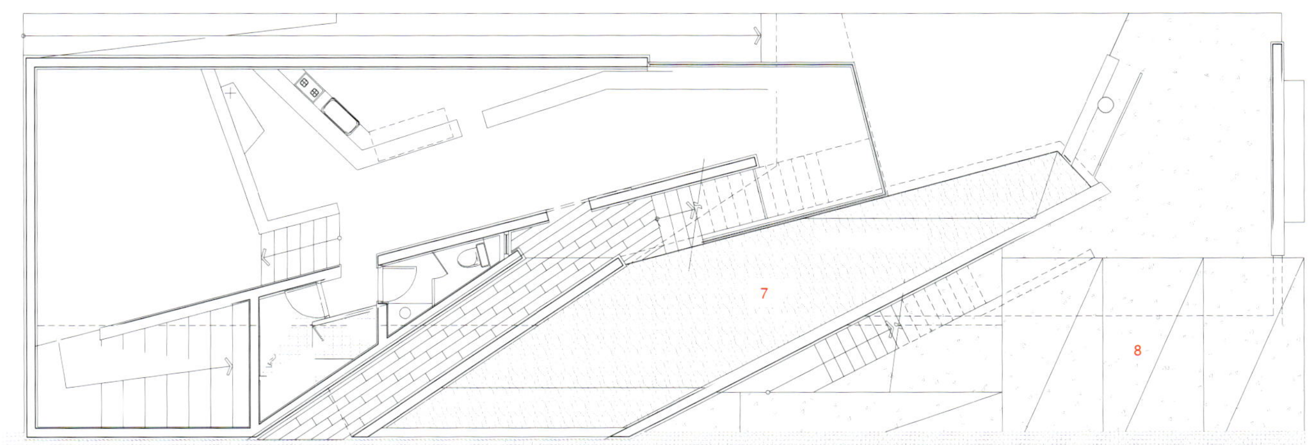

1ST FLOOR PLAN

Office on the 2nd floor Office on the 2nd floor

NEIGHBORHOOD FACILITY

MINI Building

Publisher • Editorial Director | Jung Heung Chae

Print in Korea
ISBN | 978-89-7212-118-3
Price | USA_$68 (68,000won)
Registration No. 2004-000166

10F, 15, Teheran-ro 22-gil, Gangnam-gu, Seoul, Korea
T: +82-2-538-7333
www.ancbook.com